Southern Space Studies

The Southern Space Studies series presents analyses of space trends, market evolutions, policies, strategies and regulations, as well as the related social, economic and political challenges of space-related activities in the Global South, with a particular focus on developing countries in Africa and Latin America. Obtaining inside information from emerging space-faring countries in these regions is pivotal to establish and strengthen efficient and beneficial cooperation mechanisms in the space arena, and to gain a deeper understanding of their rapidly evolving space activities. To this end, the series provides transdisciplinary information for a fruitful development of space activities in relevant countries and cooperation with established space-faring nations. It is, therefore, a reference compilation for space activities in these areas.

The volumes of the series are peer-reviewed.

Annette Froehlich
Editor

Space Fostering Latin American Societies

Developing the Latin American Continent Through Space, Part 4

Editor
Annette Froehlich
SpaceLab
University of Cape Town
Rondebosch, South Africa

ISSN 2523-3718 ISSN 2523-3726 (electronic)
Southern Space Studies
ISBN 978-3-031-20674-0 ISBN 978-3-031-20675-7 (eBook)
https://doi.org/10.1007/978-3-031-20675-7

This Springer imprint is published by the registered company Springer Nature Switzerland AG
The registered company address is: Gewerbestrasse 11, 6330 Cham, Switzerland

Contents

Space as a Tool for Development in Brazil: A Law and Development Perspective

Adriana Simões and Felipe Kotait Buchatsky

Abstract

Brazil, as many developing countries, faces several challenges regarding its development levels, whether in its social, economic, cultural, or political senses. Aiming to explore potential solutions to the complexities identified in the Brazilian society, this chapter departs from a perspective of Law and Development in order to assess such complexities under the lens of the contributions of the Brazilian legal framework on space matters to induce greater levels of development, especially on topics such as infrastructure, telecommunications, and environment, without prejudice to multidisciplinary approaches involving other areas of knowledge in addition to law. Using important indexes and data of the respective key sectors for development in Brazil, as well as the main laws and regulations on each topic, this chapter seeks to share with our peers in Latin America and around the world the Brazilian experience and the lessons learned regarding the legal discipline on the outer space and the potential contributions to national development.

A. Simões (✉) · F. K. Buchatsky
Mattos Filho, São Paulo, Brazil
e-mail: adriana.simoes@mattosfilho.com.br

F. K. Buchatsky
e-mail: felipe.buchatsky@mattosfilho.com.br

F. K. Buchatsky
São Paulo Law School of Fundação Getulio Vargas, São Paulo, Brazil

A. Froehlich (ed.), *Space Fostering Latin American Societies*, Southern Space Studies,
https://doi.org/10.1007/978-3-031-20675-7_1

1 Introduction

As one of the major concerns regarding the peaceful coexistence of different countries, the concept of sovereignty is crucial for the harmonious relationship between nations worldwide. Even though it is commonplace to determine the borders, airspace and exclusive economic zone of sovereign states, a different approach applies when it comes to debating the limits of outer space.

According to article I, second paragraph, of the Treaty on Principles Governing the Activities of States in the Exploration and Use of Outer Space, including the Moon and Other Celestial Bodies (OST) (1967), which was enacted in Brazil by Presidential Decree No. 64.362, of 17 April 1969,[1] the outer space shall be free for exploration and use by all states, with free access to all areas of celestial bodies, in an equal and non-discriminatory basis, and in accordance with international law.

Considering the opening of the space for any nation, regardless of the stage of economic and/or scientific development of the countries interested in exploring and using the outer space, as per article I, first paragraph, of the OST, the Global South has the opportunity to take advantage of the resources offered by space and space research to induce higher levels of development.

Besides the democratic spirit guiding the use and exploration of space at the international level, a domestic perspective of the potential for democratization through space must be developed. Especially in regions with chronic challenges in fighting poverty and addressing their poor infrastructure, such as Latin America, the development of the society highly depends on the ability of local governments to promote initiatives that induce development. In this effort, the space appears as an important factor to be studied by public officials and the private sector, given the opportunities of investments and innovation that enable the progress of societies and the development of nations.

In this chapter, outer space will be analyzed as an important tool for development of nations worldwide, but particularly focusing on the Brazilian scenario, which presents numerous challenges also shared by its Latin American peers. This chapter is affiliated with an optimistic view of the role of the space and its features to promote development and improve not only the economic environment of countries, but also the social concerns on the human development of societies and their access to high quality public utilities.

[1] Presidential Decree No. 64.362, 17 April 1969, www.planalto.gov.br/ccivil_03/decreto/1950-1969/D64362.html (all websites cited in this publication were last accessed and verified on 21 July 2022).

2 An Approach to Law and Development

As indicated in the introduction to this chapter, the main focus of this publication will be the analysis of the space as a tool for the development of nations, especially the Brazilian state. However, according to Kattie Willis, "development can be considered at a number of spatial scales", and "how development is defined may differ by scale".[2]

Considering the broad concept of development and its intersections, this article will adopt as a reference the field of Law and Development. From a historical perspective, academic work on Law and Development goes back to the 1960s in the United States of America, when law started to be understood as an important issue for development, especially for reform projects.[3]

Within the institutional approach that exists in the field of Law and Development, it is worth highlighting that technology is becoming a key issue in debates on the development of societies. According to Michael Trebilcock and Mariana Mota Prado, "technology and institutions act as complements, rather than substitutes in advancing progress, human well-being, and development".[4] This assumption is deeply related to the role of the space in development, a sector which is constantly evolving, relying on technological innovations to surpass current knowledge and achieve new discoveries.

Characterized by fluctuations over time, with periods of prominence and periods of decline, the Law and Development field has an approach to law not as a restrict and impermeable science, but rather as a multidisciplinary area of knowledge that can dialogue with other sciences, such as economy, political science, sociology, and anthropology. While this multidisciplinary approach is a recognition that law may not be the only solution for complex problems that plague societies, it also indicates that law is an important matter when analyzing the promotion of development.

For example, Katharina Pistor argues that the capital is made of an asset, combined with a legal code, which enables the creation of wealth.[5] Even though the author focuses on a private law perspective and on the importance of private attorneys in this process,[6] her starting point is a bigger concern on "the creeping erosion of the legitimacy of states and their laws in the face of growing inequality".[7]

---

[2] Kattie Willis, Theories and practices of development, Second Edition, Routledge, 2011, p. 8, www.ru.ac.bd/stat/wp-content/uploads/sites/25/2019/03/408_03_Willis-Theories-and-Practices-of-Development-2011.pdf.

[3] David Trubek, Academy and Law and Development, in: Feyter; Turkerlli; Moerloose, Encyclopedia of Law and Development, Edward Elgar, 2021, p. 4.

[4] Michael Trebilcock and Mariana Mota Prado, Advanced introduction to law and development, Northampton: EE, 2017, p. 45.

[5] Katharina Pistor, The code of capital: how the law creates wealth and inequality, Princeton and Oxford, 2019, p. 2.

[6] Ibid., p. 21.

[7] Ibid., p. 22.

It is based on this realization of the growing inequality in nations, especially during the coronavirus pandemic, that this chapter aims to explore the role of space law in reducing such inequality by enabling strategies for national development that will ultimately benefit society as a whole, including the most vulnerable groups.

3 Brazilian Legal Framework on Space Matters

In order to explore the challenges faced by the Brazilian society to promote higher levels of development, it is worth mentioning the state of the art of the legal framework on space-related matters. Even though there has been a recent increase in the number of Brazilian legislations and regulations on space, the importance of space law has been a part of the history of the country for decades.

In short, Brazil's legal framework regarding space is a compilation of rules in different hierarchical levels. Besides constitutional provisions, there are international conventions ratified by the Brazilian government, statutes, presidential decrees, and ordinances issued by other authorities, such as the Brazilian Space Agency (AEB, in the acronym in Portuguese), which is linked to the Presidency of the Republic and, more intimately, to the Ministry of Science, Technology and Innovation.

Therefore, the sources of Brazilian space law are sparse in the legal system, which, while denoting the importance of the theme as a state policy in Brazil, insofar as it is contemplated in the Constitution as the highest norm of the nation, also indicates that the identification of the relevant regulations requires the advice of specialists who are used to operating in the local legal system.

3.1 Constitutional Provisions

As stated in a previous work published on part 3 of this book series,[8] the Brazilian Federal Constitution of 1988 carries some dispositions on space, especially regarding the division of powers within the state, establishing that the Union has the attribution to explore aerospace navigation, as per article 21, item XII, subitem *c* of the Constitution.[9] According to Paulo Gustavo Gonet Branco, in a book co-authored by Gilmar Ferreira Mendes, a justice of the Brazilian Supreme Federal

[8] Ian Grosner, Adriana Simões and Marina Stephanie Ramos Huidobro, What is Brazil Doing to Develop Its Commercial Space Program?, in: Froehlich, A. (eds) Space Fostering Latin American Societies, Southern Space Studies, Springer, 2022, Cham, https://doi.org/10.1007/978-3-030-97959-1_1.

[9] Brazilian Federal Constitution, 5 October 1988, www.planalto.gov.br/ccivil_03/constituicao/constituicao.htm.

Court,[10] the attributions of the Union indicated in article 21 of the Constitution include subjects related to the sovereign power of the Brazilian state, or that otherwise require the attention of the central government due to security or efficiency reasons.[11]

The Constitution also provides for the attribution of the Union to legislate on matters concerning space law and aerospace navigation and defense, in its article 22, items I and XXVIII.[12] According to Alexandre de Moraes, also a Federal Supreme Court justice, article 22 of the Constitution "demonstrates a clear supremacy [of the Union] in relation to the other federative bodies [such as municipalities and states], due to the relevance of such provisions".[13] In another remark, Gilmar Ferreira Mendes and Paulo Gustavo Gonet Branco affirm that "the most relevant and common interest subjects of the social life in the country in its numerous faraway locations are listed in the catalog of article 22 of the Federal Constitution".[14]

3.2 International Conventions and Treaties

In terms of international conventions and treaties, Brazil has ratified four out of the five United Nations agreements related to activities in outer space. Besides the OST (1967), as previously mentioned, the Agreement on the Rescue of Astronauts, the Return of Astronauts and the Return of Objects Launched into Outer Space (ARRA) (1969) has been incorporated in the Brazilian legal system by means of the Presidential Decree No. 71.989, of 26 March 1973.[15] Likewise, the Convention on International Liability for Damage Caused by Space Objects (LIAB) (1972) has been ratified by the Presidential Decree No. 71.981, of 22 March 1973[16] and, decades later, the Convention on Registration of Objects Launched into Outer Space (REG) (1975) was ratified by the Presidential Decree No. 5.806, of 19 June 2006.[17]

[10] The Brazilian Supreme Federal Court is responsible for the guard of the Constitution as has attributions related to the abstract control of constitutionality, among others.

[11] Gilmar Ferreira Mendes and Paulo Gustavo Gonet Branco, Curso de direito constitucional (Série IDP. Linha doutrina), p. 452, Minha Biblioteca, (17th edition), Editora Saraiva, 2022.

[12] Brazilian Federal Constitution, 5 October 1988, www.planalto.gov.br/ccivil_03/constituicao/constituicao.htm.

[13] Alexandre de Moraes, Direito Constitucional, p. 375, Minha Biblioteca (38th edition), Grupo GEN, 2022.

[14] Mendes and Branco, Curso de direito constitucional, p. 452.

[15] Presidential Decree No. 71.989, 26 March 1973, www.planalto.gov.br/ccivil_03/decreto/1970-1979/D71989.html.

[16] Presidential Decree No. 71.981, 22 March 1973, www.planalto.gov.br/ccivil_03/decreto/1970-1979/D71981.html.

[17] Presidential Decree No. 5.806, 19 June 2006, www.planalto.gov.br/ccivil_03/_ato2004-2006/2006/decreto/D5806.htm.

Even though historical documents made available by the United Nations show that the draft resolution that adopted the Agreement Governing the Activities of States on the Moon and Other Celestial Bodies (MOON) (1979) had Brazil as one of its authors,[18] a table also made available by the United Nations indicating the status of such international agreements, as of 1 January 2022, shows that Brazil has neither signed nor ratified the MOON (1979).[19]

3.3 Federal Laws

Besides constitutional provisions and international law rules, the Brazilian legal framework for space matters includes federal laws enacted by the National Congress. Some of these laws have a direct impact in the space sector, since they establish means for scientific development through the adoption of incentives to scientific research, innovation, technology, and professional training, such as Federal Law No. 10.973, of 2 December 2004.[20]

As a specific rule for space matters, we may highlight Federal Law No. 8.854, of 10 February 1994, which created the AEB, a federal autarchy of civil nature that aims to promote the development of space activities of Brazil's national interest.[21] Besides its administrative and financial independence, which are important to achieve its goals, AEB may act directly or indirectly via contracts, conventions and adjustments in Brazil and abroad. For international agreements or conventions, the signature shall occur in coordination with the Ministry of Foreign Affairs.

3.4 Presidential Decrees

Presidential decrees are a simpler way to enact rules on a federal level, since this category prescinds approval from the National Congress, even though the President of the Republic should only enact decrees in specific circumstances provided by the Federal Constitution, such as the faithful enforcement of laws.[22] The Brazilian legal framework carries several presidential decrees providing on the subject.

---

[18] United Nations Digital library, Agreement governing the activities of States on the moon and other celestial bodies: draft resolution, https://digitallibrary.un.org/record/9498.

[19] United Nations Committee on the Peaceful Uses of Outer Space, Status of International Agreements relating to activities in outer space as at 1 January 2022, www.unoosa.org/res/oosadoc/data/documents/2022/aac_105c_22022crp/aac_105c_22022crp_10_0_html/AAC105_C2_2022_CRP10E.pdf, p. 5.

[20] Federal Law No. 10.973, 2 December 2004, www.planalto.gov.br/ccivil_03/_ato2004-2006/2004/lei/l10.973.htm.

[21] Federal Law No. 8.854, 10 February 1994, www.planalto.gov.br/ccivil_03/Leis/L8854.htm.

[22] Brazilian Federal Constitution, 5 October 1988, www.planalto.gov.br/ccivil_03/constituicao/constituicao.htm.

Prior to the redemocratization of Brazil, Presidential Decree No. 88.136, of 1 March 1983, created the Alcântara Launch Center (CLA, in the acronym in Portuguese), under the former Ministry of Aeronautics, aiming to execute and support launch activities, track aerospace devices, and execute tests and experiments.[23]

Federal Law No. 8.854, of 10 February 1994, established that the update of the National Policy for Development of Space Activities (PNDAE, in the acronym in Portuguese) was an attribution of the AEB, and that the Brazilian space activities should be organized systematically, as established by the executive branch of government. In order to guarantee the faithful enforcement of both provisions, the President of the Republic enacted Presidential Decrees No. 1.332, of 8 December 1994,[24] and No. 1.953, of 10 July 1996.[25]

Presidential Decree No. 1.332, of 8 December 1994, establishes the update of the PNDAE with goals and guidelines that shall drive the actions of the Brazilian government related to the promotion of development of space activities of national interest. In turn, Presidential Decree No. 1.953, of 10 July 1996, establishes the National System for the Development of Space Activities (SINDAE, in the acronym in Portuguese) to organize the performance of activities seeking national interest space development.

More recently, the presidential decree has been used as tool not exactly related to the faithful enforcement of laws, but rather for enacting general rules regarding space matters. For example, Presidential Decree No. 9.839, of 14 June 2019,[26] establishes a regulation of the Brazilian Space Program Development Committee (CDPEB, in the acronym in Portuguese), which is a body formed by certain Ministries of the executive branch of government and the Attorney-General that advises the President of the Republic in formulating proposals on a few subjects related to the space, such as subsidies to potentialize the Brazilian Space Program.

Also, Presidential Decree No. 10.220, of 5 February 2020, has been used by the President of the Republic to promulgate the Technological Safeguards Agreement signed in Washington D.C. by and between the Federative Republic of Brazil and the United States of America, duly approved by the Brazilian National Congress in 2019. The agreement relates to the participation of the United States of America in space launches from the Alcântara Space Center[27] (CEA, in the acronym

---

[23] Presidential Decree No. 88.136, 1 March 1983, www2.camara.leg.br/legin/fed/decret/1980-1987/decreto-88136-1-marco-1983-438606-publicacaooriginal-1-pe.html.

[24] Presidential Decree No. 1.332, 8 December 1994, www.planalto.gov.br/ccivil_03/decreto/1990-1994/d1332.htm.

[25] Presidential Decree No. 1.953, 10 July 1996, www.planalto.gov.br/ccivil_03/decreto/1996/d1953.htm.

[26] Presidential Decree No. 9.839, 14 June 2019, www.planalto.gov.br/ccivil_03/_Ato2019-2022/2019/Decreto/D9839.htm#art10.

[27] Grosner, Simões and Huidobro, "What is Brazil Doing to Develop Its Commercial Space Program?", p. 3, it is important to clarify that, while the CEA is a project under implementation, the CLA is actually a Brazilian Air Force unit that coordinates the site located in the city of Alcântara.

in Portuguese) and seeks to prevent unauthorized access or transfer of technologies related to certain launches from the CEA that involve U.S. launch vehicles or launch vehicles that include or carry equipment authorized for export by the government of the United States of America.

3.5 Ordinances as Technical Norms

Lastly, the Brazilian legal framework related to space matters also includes norms of a more technical nature, rather than political. Recently, the AEB has published in the Federal Official Gazette two parts of the Brazilian Space Regulation (REB, in the acronym in Portuguese), by means of Ordinance No. 698, of 31 August 2021.[28] The published REBs bring technical procedures that shall apply for the operator's license[29] to execute space launch activities in the Brazilian territory, as well as for the concession of the launch authorization[30] by the AEB.

It is worth mentioning that the regulatory framework with technical standards is constantly being improved: in April 2022, the AEB submitted to public consultation a proposal for a third part of the REB, in order to establish certain rules for insurance requirements for space launches.[31] As a result, Ordinance No. 1.019, of 23 December 2022, has turned the proposal under public consultation into legal norm and created REB Part 03, providing the guidelines, conditions, and baseline amounts for the hiring of insurances for space launches from Brazilian territory.[32] It is also important to highlight that the public consultation of the Brazilian society regarding technical aspects of space activities is an indicator of the current priorities of the government and the intention of Brazilian public authorities to establish adequate regulations and dialogue with experts on the subject.

3.6 A General Space Law?

Despite the sparse norms described above, some authors point out to the need of a Brazilian General Space Law, in line with the assumptions made by Brazil under international treaties and conventions. Márcia Alvarenga dos Santos, Petrônio Noronha de Souza, and Ian Grosner, after exposing a comparative study on the

[28] Ordinance No. 698, 31 August 2021, www.in.gov.br/web/dou/-/portaria-n-698-de-31-de-agosto-de-2021-341897559.

[29] REB Part 01, www.gov.br/aeb/pt-br/servicos/licenciamento/copy_of_REB_Parte01LicenadeOperadoragosto2021.pdf.

[30] REB Part 02, www.gov.br/aeb/pt-br/servicos/licenciamento/REB_Parte02AutorizaodeLanamentoFinalagosto2021.pdf.

[31] Public consultation draft of REB Part 03, www.gov.br/participamaisbrasil/regulamento-espacial-brasileiro-requisitos-de-seguro-para-lancamento-espacial.

[32] Ordinance No. 1.019, 23 December 2022, https://www.in.gov.br/web/dou/-/portaria-n-1.019-de-23-de-dezembro-de-2022-454136749.

experience of other countries, such as Finland, Portugal, Argentina and the United States of America, indicate that it is expected from a national space legislation the legal certainty and modernity needed to "promote competitiveness and secure and sustainable growth to the space industry, allowing, thus, to attract new actors and investments to Brazil".[33]

According to the authors, "if it is the duty of the State Party to the OST (1967) to authorize and supervise the space activities of its nationals, whether they are governmental or not, one of the main objectives of the Brazilian General Space Law should be to regulate the activities subject to this duty and also those for which Brazil will be held liable, pursuant to art. VII of the OST (1967)".[34] The scope of the Brazilian General Space Law, in turn, would be "to regulate space activities that are subject to the authorization and supervision of the Brazilian state, as per art. VI of the OST (1967), and those by which Brazil will be liable, as provided for in the LIAB (1972)".[35]

4 Main Challenges to the Development of Brazil

The current challenges of Brazilian society seem chronic and difficult to resolve. Besides the worldwide effects of the Covid-19 pandemic, which still affects countries in different ways, Brazil has been experiencing economic struggles. Brazil is the 13th biggest economy in the world,[36] in terms of gross domestic product for 2021 in American dollars, in comparison to the seventh position occupied by the country in 2014.[37]

According to data updated by the Organization for Economic Co-operation and Development (OECD) in June 2022, the Brazilian gross domestic product growth shall be 0,6% in 2022, due to rise in inflation, the war in Ukraine and tougher financial conditions.[38]

Certainly, the economic crisis deepens the inequality in social relations in Brazil, aggravating non-economic indexes, such as starvation, unemployment, and poverty. Impacts may also be observed in other sectors that depend on huge investments and that, in periods of economic crisis, fight with other sectors for a piece of the public budget and depend on the economic attractiveness of the country to obtain private resources, including foreign ones.

[33] Márcia Alvarenga dos Santos, Petrônio Noronha de Souza and Ian Grosner, A necessidade de uma Lei Geral do Espaço no Brasil, Revista de Direito da Universidade de Brasília, 2020, p. 122, v. 4, n. 3, pp. 106–138, https://periodicos.unb.br/index.php/revistadedireitounb/article/view/34672.

[34] Ibid., p. 123.

[35] Ibid., p. 124.

[36] Country Economy, Comparison: Annual GDP 2021, https://countryeconomy.com/gdp?year=2021.

[37] Country Economy, Comparison: Annual GDP 2014, https://countryeconomy.com/gdp?year=2014.

[38] Organization for Economic Co-operation and Development, Brazil Economic Snapshot, https://www.oecd.org/economy/brazil-economic-snapshot/.

Despite the challenging scenario currently faced in Brazil, not only in terms of economic growth, but also in term of reduction of inequalities, it is necessary to recognize that some innovative initiatives contribute to inducing higher levels of development in the different sectors of society. This chapter focuses on this view of what has been done in Brazil and what is planned for the future, as an attempt to disseminate experiences that can be replicated or improved in the country or elsewhere.

5 Space as a Tool for Development: Remedying the Brazilian Complexities

Usually, writing and talking about development involves a debate on macroeconomic factors, new legislations, income transfer programs, social benefits, heavy investments in the industry and other sectors, and privatization of certain activities. Certainly, these aspects are crucial to address the complexities of a society by means of organized public policies. However, restricting the debate to issues of great magnitude overshadows the role of other matters that, rather than replacing robust and deep reforms, may walk side by side and complement a larger and more complex project of reform.

In the subsections below, this chapter will explore the state of the art of effective and potential contributions of the space sector with the development in major areas that face important complexities in Brazil. After exposing some of the current concrete challenges in Brazil, we will indicate Brazilian experiences of using the space as a tool for inducing greater levels of development in five different delicate areas of Brazilian society: (i) telecommunications, (ii) education and culture, (iii) improvement of the infrastructure, (iv) environmental protection, and (v) debureaucratization.

5.1 Telecommunications

The telecommunication sector is crucial in every country. In Brazil, this cruciality is reiterated by the need to connect distant places, considering Brazil's continental dimensions. Like other sectors, such as the aviation industry, the telecommunication sector is subject to several regulations, many of them issued by a regulatory agency specifically created with the function to watch over telecommunications. The National Telecommunications Agency (ANATEL, in the acronym in Portuguese) is an autarchy under a special regime, integrating the federal public administration, but characterized by administrative independence, absence of hierarchical subordination, and financial autonomy, to best exercise its attributions.[39]

[39] Federal Law No. 9.472, 16 July 1997, www.planalto.gov.br/ccivil_03/leis/l9472.htm#:~:text=LEI%20N%C2%BA%209.472%2C%20DE%2016%20DE%20JULHO%20DE%201997.&text=

As a general balance of the telecommunication infrastructure in Brazil, it can be argued that "the broadband service in the country is expensive and the connection speed is low", figuring in the 79th position among 148 countries ranked by the State of Internet Report 2017, in terms of internet average speed.[40] Over the past few years, the competitiveness among phone carriers reduced the average price of the phone and fixed broadband services, even though the tax burden is still identified as a key deterrent for the expansion of networks to remote regions of Brazil.[41]

In this regard, the space plays an important role in providing telecommunication services, which is recognized by the federal law that created ANATEL. According to Federal Law No. 9.472, of 16 July 1997, ANATEL has the attribution to provide for the specific requirements for the execution of telecommunication services that use satellite, geostationary or not, accessed from the national territory or from abroad.[42]

Even though Federal Law No. 9.472, of 16 July 1997, establishes a differentiation between Brazilian and foreign satellites, both are admitted under the legislation.[43] Brazilian satellites[44] (i.e. those that use orbit and radio spectrum resources notified by the country, or distributed or assigned to it, and whose control and monitoring station is installed in Brazilian territory) shall have preference in case it provides equivalent conditions to those satellites of third parties.[45] Foreign satellites,[46] in turn, are allowed when contracted with a company incorporated under Brazilian law and with headquarters and administration in Brazil, acting as legal representative of the foreign operator.[47]

Also according to Federal Law No. 9.472, of 16 July 1997, the right to exploit a Brazilian satellite to transport telecommunication signals ensures the occupation of the orbit and the use of radio frequencies intended to control and monitor the

Disp%C3%B5e%20sobre%20a%20organiza%C3%A7%C3%A3o%20dos,Constitucional%20n%C2%BA%208%2C%20de%201995.

[40] Infraestrutura - O que é, quais os tipos, seus desafios e prioridades, Portal da Indústria, www.portaldaindustria.com.br/industria-de-a-z/infraestrutura/.

[41] Circe Bonatelli, Brasil melhora em ranking de preços de telefonia móvel e banda larga, diz Anatel, CNN Brasil, 22 March 2021, www.cnnbrasil.com.br/business/brasil-melhora-em-ranking-de-precos-de-telefonia-movel-e-banda-larga-diz-anatel/.

[42] Federal Law No. 9.472, 16 July 1997, www.planalto.gov.br/ccivil_03/leis/l9472.htm#:~:text=LEI%20N%C2%BA%209.472%2C%20DE%2016%20DE%20JULHO%20DE%201997.&text=Disp%C3%B5e%20sobre%20a%20organiza%C3%A7%C3%A3o%20dos,Constitucional%20n%C2%BA%208%2C%20de%201995.

[43] Ibid.

[44] According to ANATEL Resolution No. 748, 22 October 2021, a Brazilian satellite is the one that uses orbit resources and radio spectrum resources notified by Brazil to the International Telecommunication Union (ITU) and whose control and monitoring station is installed in Brazil.

[45] Ibid.

[46] According to ANATEL Resolution No. 748, 22 October 2021, a foreign satellite is the one that uses orbit and radio spectrum resources notified by other countries to the International Telecommunication Union (ITU).

[47] Ibid.

satellite and the telecommunication via satellite, for an extendable period of up to 15 years.[48] Conducting telecommunication activities without the relevant authorization for exploitation of satellite is a clandestine practice, subject to criminal penalties of imprisonment and fine.[49]

To further detail the succinct procedures required by Federal Law No. 9.472, of 16 July 1997, ANATEL enacted its Resolution No. 748, of 22 October 2021, approving the General Regulation of Satellites Exploitation: in short, such regulation provides general conditions for the exploitation of satellites on Brazilian territory and the granting of satellites exploitation rights for Brazilian or foreign satellites.[50] For clarification purposes, the exploitation right, which will be further mentioned in this chapter, refers to an administrative act authorizing the use of orbit and radiofrequency resources to control and monitor the satellite, the satellite telecommunication and the offering of a satellite infrastructure for the traffic of any kind of telecommunication signals.[51]

The General Regulation of Satellites Exploitation establishes some requirements for the obtainment of satellites exploitation rights, such as: (i) be a legal entity of private or public law incorporated under Brazilian laws and headquartered and administered in Brazil, (ii) not be prevented from participating in bidding procedures or contracting with the public power, (iii) have legal and technical qualification to exploit a satellite, (iv) have economic-financial capacity and be in compliance with certain tax and social security requirements, (v) present a simplified technical design of the satellite communication system, as well as a declaration of compliance with the applicable regulations and awareness of the conditions granted.[52]

Further requirements may apply depending on the category of the satellite (i.e. Brazilian or foreign), however, in both cases, a prefixed payment is due in order to obtain the exploitation right or its renewal, correspondent to the amount of BRL 102.677,00 (around US$ 19.000,00, according to the currency exchange rate for mid-July 2022).[53] The maximum period for entry into operation of geostationary satellites will be five years for Brazilian satellites and two years for foreign ones, counted as of the publication in the Federal Official Gazette of the extract of the act issued by ANATEL granting the satellites exploitation rights, while non-geostationary satellites have their period and conditions for entry into operation established case by case.[54] The exploitation right extinguishes in its final

[48] Ibid.
[49] Ibid.
[50] ANATEL Resolution No. 748, 22 October 2021, https://informacoes.anatel.gov.br/legislacao/resolucoes/2021/1595-resolucao-748.
[51] Ibid.
[52] Ibid.
[53] Ibid.
[54] Ibid.

term, termination of the satellite's lifespan or upon annulment, waiver, bilateral termination, among other hypotheses.[55]

The application of the legal discipline on satellites for telecommunication services had some recent important outcomes. Over the past few months, ANATEL has approved the operation of several satellites of major companies, such as:

- exploitation right and use authorization of radiofrequency for satellite IS-901 by Intelsat, until 28 February 2025[56];
- use of the Starlink/SpaceX network of interconnected satellites in telecommunication operations in Brazil, until March 2027[57];
- exploitation of non-geostationary satellite system, based on a constellation of 150 satellites, by Swarm Technologies, until September 2035[58];
- exploitation right of foreign non-geostationary satellites system by Kepler Communications, until 20 January 2037, operating with up to 175 satellites[59];
- exploitation right of a constellation of non-geostationary satellites by Telesat for 293 equipment, until March 2036; and[60]
- exploitation right of OneWeb non-geostationary satellites system by OneWeb Limited/Worldvu Satellites for 774 equipment, until July 2037.[61]

Therefore, the current advances on the telecommunication sector regarding satellites, such as the exploitation authorizations issued by ANATEL over the past few months and the approval of the General Regulation of Satellites Exploitation, show that the regulation of outer space activities is a key topic in the national agenda of the sector, seeking to further develop the telecommunications in Brazil, due to its latent relevance.

[55] Ibid.

[56] Gabriela do Vale. ANATEL autoriza Intelsat a explorar satélite até 2025, Tele Síntese, 21 December 2021, www.telesintese.com.br/anatel-autoriza-intelsat-a-explorar-satelite-ate-2025/.

[57] Anatel aprova operações da rede de satélites Starlink, da SpaceX, Agência Brasil, 28 January 2022, https://agenciabrasil.ebc.com.br/geral/noticia/2022-01/anatel-aprova-operacoes-da-rede-de-satelites-starlink-da-spacex.

[58] Lúcia Berbert. ANATEL autoriza exploração de satélites LEO pela Swarm, Tele Síntese, 11 February 2022, www.telesintese.com.br/anatel-autoriza-exploracao-de-satelites-leo-pela-swarm/.

[59] Lúcia Berbert. ANATEL autoriza operação do sistema de satélites da Kepler Communications, Tele Síntese, 24 March 2022, www.telesintese.com.br/anatel-autoriza-operacao-do-sistema-de-satelites-da-klepper/.

[60] Lúcia Berbert. ANATEL autoriza constelação de satélite da Telesat até 2036, Tele Síntese, 8 April 2022, www.telesintese.com.br/anatel-autoriza-constelacao-de-satelite-da-telesat-ate-2036/.

[61] Lúcia Berbert. ANATEL dá aval ao sistema de satélites da OneWeb, Tele Síntese, 18 July 2022, www.telesintese.com.br/anatel-da-aval-ao-sistema-de-satelites-da-oneweb/.

5.2 Education and Culture

In the education sector, Brazil faces difficulties regarding the access to and quality of the public services. According to the Programme for International Student Assessment 2018 (PISA 2018), one of the key findings was that "students in Brazil scored lower than the OECD average in reading, mathematics and science [...], and 43% of students scored below the minimum level of proficiency (Level 2) in all three subjects".[62] In terms of competitiveness in 2021, CNN Brasil indicated that a study carried out by the World Competitiveness Center, of the International Institute for Management Development (IMD), classified Brazil in the last position regarding education among 64 analyzed countries.[63]

The Covid-19 pandemic contributed to an unsatisfactory performance of Brazilian education, in terms of school dropout: the press informed that, according to the school census carried out by the Brazilian National Institute of Educational Studies and Research (INEP, in the acronym in Portuguese), the dropout of 2,3% of high school students in 2020 has risen to a dropout of 5% in 2021.[64]

Especially in public schools, the lack of access to technology and internet may be identified as one of the reasons for the school dropout during the pandemic, since the distance learning became the only alternative to stop the transmission of the coronavirus and continue the studies of children and adolescents worldwide.

According to data of 2019 published by the Brazilian Institute of Geography and Statistics (IBGE, in the acronym in Portuguese), the reason why students did not use the internet was related to the cost of the internet service and of the equipment necessary to access the internet.[65] Also, the access to the mobile phones for personal use among public schools students (64,8%) is lower than the access identified for private schools students (92,6%).[66]

In this regard, the telecommunication sector may indirectly contribute with the education indexes in Brazil, considering that internet is nowadays essential to access knowledge, information and, as Covid-19 has proven to the world, also to access the classrooms. As an example of a recent initiative, during its visit to

[62] Organization for Economic Co-operation and Development, Programme for International Student Assessment (PISA) Results from PISA 2018, www.oecd.org/pisa/publications/PISA2018_CN_BRA.pdf.

[63] Rodrigo Maia, Thais Herédia and Larissa Coelho, Educação brasileira está em último lugar em ranking de competitividade, CNN Brasil, 17 June 2021, www.cnnbrasil.com.br/nacional/educacao-brasileira-esta-em-ultimo-lugar-em-ranking-de-competitividade/.

[64] Censo Escolar confirma impacto negativo da pandemia na educação básica, Globo, 20 May 2022, https://g1.globo.com/jornal-nacional/noticia/2022/05/20/censo-escolar-confirma-impacto-negativo-da-pandemia-na-educacao-basica.ghtml.

[65] IBGE—Pesquisa Nacional por Amostra de Domicílios Contínua, Acesso à Internet e à televisão e posse de telefone móvel celular para uso pessoal 2019, p. 10, https://biblioteca.ibge.gov.br/visualizacao/livros/liv101794_informativo.pdf.

[66] Ibid, p. 11.

Brazil, the founder of Starlink, Elon Musk, has announced a project to operate satellites that will enable the connection of 19.000 schools currently disconnected in rural areas of Brazil to the network.[67]

A different contribution of the space sector to education has been recently disclosed in the Federal Official Gazette: on 24 May 2022, an agreement has been entered into by and between the CLA and the Federal University of Maranhão (UFMA, in the acronym in Portuguese) to provide a mandatory internship for undergraduate students of the course of Aerospace Engineering and Interdisciplinary Bachelor of Science and Technology.[68] Aiming to complement the practical learning, as well as the technical-cultural, scientific and human relationship improvement of the students, the agreement is valid for five years and seeks to induce their participation in real situations.[69]

Lastly, it is important to highlight that Brazil has been announced as the host country for the 35th annual session of the Space Studies Program (SSP) in 2023, carried out by the International Space University (ISU).[70] The 2023 edition will take place in the city of São José dos Campos, in the state of São Paulo, a city which has been classified by its mayor as "the Brazilian hub of the aerospace cluster".[71] According to the ISU, the National Institute for Space Research (INPE, in the acronym in Portuguese) and the Aeronautics Institute of Technology (ITA, in the acronym in Portuguese) are two partners that will host the SSP 2023, supported by the city of São José dos Campos and the AEB.[72]

Besides the educational perspective, culture is a crucial factor to be analyzed in Brazilian society. According to the System of Information and Cultural Indicators of the IBGE, the average consumption expenses of Brazilian families in 2017–2018 corresponded to approximately BRL 4.100,00 per month, of which BRL 428 291,18 (around US$ 75,00, according to the currency exchange rate for late 2018) were dedicated to expenses with culture (i.e. approximately 7%), considering an average sized family of three individuals.[73] Such amounts may vary according to race, gender age, and level of schooling.

---

[67] Naty Falla, Elon Musk anuncia Starlink para conectar 19 mil escolas e monitorar Amazônia, Forbes, 20 May 2022, https://forbes.com.br/forbes-money/2022/05/elon-musk-anuncia-starlink-para-conectar-19-mil-escolas-e-monitorar-amazonia/.

[68] Extrato de Convênio, 2 June 2022, www.in.gov.br/web/dou/-/extrato-de-convenio-404937231.

[69] Ibid.

[70] International Space University, International Space University's 35th Space Studies Program to Convene in Brazil in 2023, 1 July 2022, www.isunet.edu/international-space-universitys-35th-space-studies-program-to-convene-in-brazil-in-2023/.

[71] Ibid.

[72] Ibid.

[73] Data available in section "*Gastos das famílias*", at www.ibge.gov.br/estatisticas/multidominio/cultura-recreacao-e-esporte/9388-indicadores-culturais.html?=&t=resultados.

In order to preserve the Brazilian culture in the aviation and aerospace sectors, we may highlight the Aerospace Museum, linked to the Aeronautics Historical-Cultural Institute of the Brazilian Air Force, located in Rio de Janeiro, which has a virtual tour under construction.[74] In São José dos Campos, city in the state of São Paulo that hosts the Brazilian aircraft manufacturer Embraer, the Brazilian Aerospace Memorial, built by the Aerospace Technology General-Command, has an outdoor exhibition with replicas of the Brazilian Space Program rockets, such as a full scale satellite launch vehicle.[75] Both institutions have free admission for visitors.

Also in the state of São Paulo, the city of Campinas hosts the Open Museum of Astronomy with activities for children's school excursions, offering a simulation of a space program, in which the students build their own rockets with recyclable materials and launch these rockets using water and air as fuels, reaching over 30m in altitude.[76] In the city of São Paulo, two temporary exhibitions called Space Future and Space Adventure took place in 2021 and brought to Brazil items related to previous and future missions of the National Aeronautics and Space Administration (NASA).[77]

5.3 Improvement of the Infrastructure

Although the infrastructure of a certain country is a key issue in nations worldwide, Brazil has some particularities that demand vast investments in different kinds of infrastructure, especially considering its continental dimensions and the need to build roads, airports, bridges, railroads, subways, as well as to provide sanitation, street lighting, logistics, among other relevant services.

According to data of the Brazilian Association of Infrastructure and Basic Industries (ABDIB, in the acronym in Portuguese) (Fig. 1),[78] the investment in the sector in 2019 reached BRL 123,9 billion (around US$ 30,9 billion, according to the currency exchange rate for late 2019), which is less than half of the

[74] Força Aérea Brasileira, Museu Aeroespacial – Instituto Histórico-Cultural da Aeronáutica, www2.fab.mil.br/musal/.

[75] São José dos Campos, MAB – Memorial Aeroespacial Brasileiro, 20 January 2022, https://sjc.com.br/2022/01/20/mab-memorial-aeroespacial-brasileiro/.

[76] Museu Aberto de Astronomia, Oficina Lançamento de Foguetes, https://museuabertodeastronomia.com.br/oficina-lancamento-de-foguetes/.

[77] Camila Mazzotto, Exposições inéditas sobre missões da Nasa chegam a SP em julho e agosto, Revista Galileu, 21 July 2021, https://revistagalileu.globo.com/Cultura/noticia/2021/07/exposicoes-ineditas-sobre-missoes-da-nasa-chegam-sp-em-julho-e-agosto.html.

[78] Bluebook Infrastructure: A radiography of infrastructure projects in Brazil, Brazilian Association of Infrastructure and Basic Industries, p. 5, www.abdib.org.br/wp-content/uploads/2021/05/BLUEBOOK_INFRASTRUCTURE.pdf.

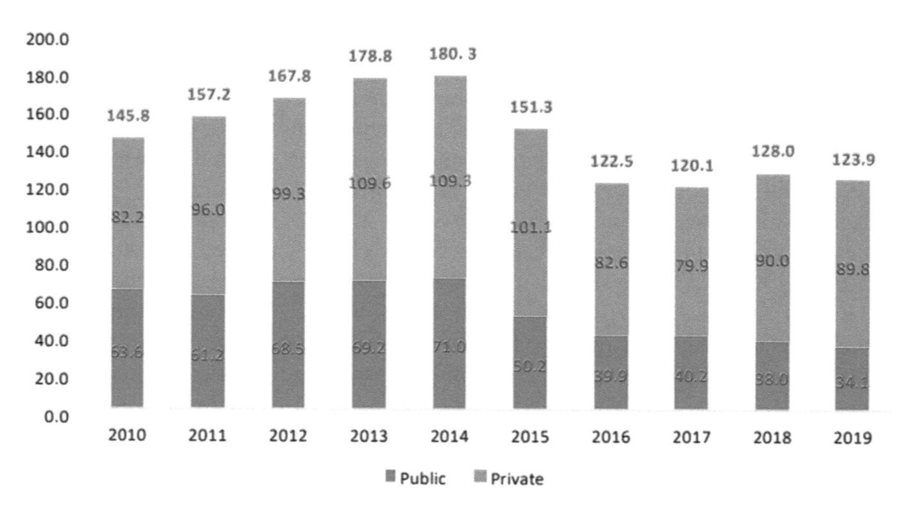

Fig. 1 Evolution of investments in Infrastructure (in billion BRL per year, constant values at 2019 prices, excluding the oil and gas sector) (*Source* ABDIB)

annual investment needed for ten years to reduce the main issues related to the socioeconomic development of the country.[79]

A certain mechanism provided by the Brazilian administrative law has been frequently used, for example, in the airport sector to improve the quality of the infrastructure: concession agreements entered into by and between the Union, represented by the Brazilian National Civil Aviation Agency (ANAC, in the acronym in Portuguese), and private concessionaires that participate in a bidding process. Since 2011, more than 40 airports have been transferred from the Brazilian state-owned company responsible for the administration of airports to private parties that shall follow a schedule of the works to improve the infrastructure and reach stricter levels of quality of the services provided to passengers.

In the space sector, a different mechanism has been used to stimulate the improvement of the infrastructure: two public calls have been convened in 2020 and 2021, respectively, aiming to identify companies interested in carrying out orbital launch operations for non-military space vehicles from Brazilian territory, using the CEA.[80]

According to the public calls, the objective was to make Union's goods and services available to national or foreign legal entities, for the operationalization of

[79] Abdib: Investimento em infraestrutura no Brasil é menos da metade do necessário, CNN Brasil, 1 December 2020, www.cnnbrasil.com.br/business/abdib-investimento-em-infraestrutura-no-brasil-e-menos-da-metade-do-necessario/.

[80] Public Call I: www.gov.br/aeb/pt-br/programa-espacial-brasileiro/chamamento-publico-public-call/SEI_AEB0072186EditalPTBR.pdf.

Public Call II: www.gov.br/aeb/pt-br/programa-espacial-brasileiro/chamamento-publico-public-call/chamamento-publico-1/chamamento-publico-2.pdf.

the orbital launch of the space vehicles, each public call referring to a different area of the site.[81] The participation in the public calls involved the execution of a non-disclosure agreement with the AEB and the obtainment of the operator's license as a condition precedent for the beginning of the contractual negotiation, which shall result in a contract to be entered into by and between the selected legal entity and the Aeronautical Command (COMAER, in the acronym in Portuguese).[82]

As a result of the first public call, carried out in 2020, the following four companies (of which three are American and one is Canadian) have been announced by the Brazilian Air Force (FAB, in the acronym in Portuguese), assuming different attributions under the project: Hyperion, Orion AST, C6 Launch, and Virgin Orbit.[83] Even though the public calls have been only the first step towards the effective use of the infrastructure with noble purposes, the presence of major companies in Brazilian territory, carrying out space activities in the country, is an indicator of the attractiveness of Brazil and a key factor for the improvement of the infrastructure, also taking into account the potential development of the local community and the creation of jobs for local citizens.

5.4 Environmental Protection

Brazil is worldwide known for its biodiversity and the global importance of its natural resources. According to official data of the National System of Forest Information (SNIF, in the acronym in Portuguese), the Brazilian forest area corresponded to more than 57% of its territory in 2018 (i.e. more than 4,8 million km^2).[84] In comparison with its peers, Brazil has the second biggest forest area in the world, only after the Russian Federation.[85]

According to recent data of the INPE, the deforestation rate in the Brazilian Legal Amazon verified by satellite images was 13.235 km^2 between 1 August

[81] Ibid.

[82] Ibid.

[83] FAB divulga empresas selecionadas para operação no Centro Espacial de Alcântara, Agência Força Aérea, 28 April 2021, www.fab.mil.br/noticias/mostra/37237/ESPACIAL%20-%20FAB%20divulga%20empresas%20selecionadas%20para%20opera%C3%A7%C3%A3o%20no%20Centro%20Espacial%20de%20Alc%C3%A2ntara.

[84] Serviço Florestal Brasileiro – Sistema Nacional de Informações Florestais, Painel Interativo – Área de Floresta Natural – Brasil, https://snif.florestal.gov.br/pt-br/os-biomas-e-suas-florestas#:~:text=A%20%C3%A1rea%20de%20floresta%20do,apenas%202%25%20s%C3%A3o%20florestas%20plantadas.

[85] Brasil possui a 2ª maior área de florestas do mundo, Globo, 21 March 2020, https://g1.globo.com/sp/campinas-regiao/terra-da-gente/noticia/2020/03/21/brasil-possui-a-2a-maior-area-de-florestas-do-mundo.ghtml.

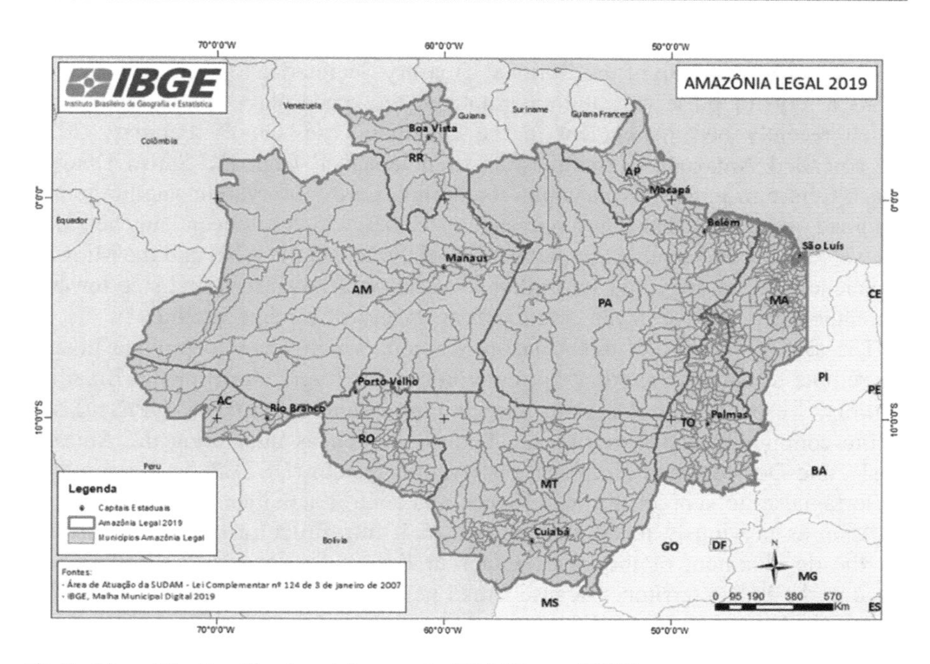

Fig. 2 Map of the Brazilian Legal Amazon in 2019 (*Source* IBGE)

2020 and 31 July 2021, which represents an increase of more than 20% when compared to the previous period.[86]

For reference purposes, the Legal Amazon occupies more than five million km^2 of the Brazilian territory identified in the map (Fig. 2)[87] and is a zone in which every rural real estate, as a rule, must maintain an area with native vegetation coverage of at least (i) 80% in real estate located in forest area, (ii) 35% in real estate in the Brazilian *cerrado* (savannah) area, or (iii) 20% in real estate located in an area of general fields, as compared to the general rule of 20% of maintenance of area with native vegetation coverage applicable in other regions of the country.[88]

In terms of investment in environmental protection, the resources invested in the sector in 2021 reached the lowest level over the last twelve years, equivalent to BRL 3,87 billion (around US$ 694 million, according to the currency exchange

---

[86] Inpe: desmatamento na Amazônia Legal tem aumento de 21.97% em 2021, Agência Brasil, 18 November 2021, https://agenciabrasil.ebc.com.br/geral/noticia/2021-11/desmatamento-na-amazonia-legal-tem-aumento-de-2197-em-2021.

[87] Agência IBGE Notícias, IBGE atualiza Mapa da Amazônia Legal, 29 June 2020, https://censoagro2017.ibge.gov.br/agencia-sala-de-imprensa/2013-agencia-de-noticias/releases/28089-ibge-atualiza-mapa-da-amazonia-legal.

[88] Federal Law No. 12.651, 25 May 2012, www.planalto.gov.br/ccivil_03/_ato2011-2014/2012/lei/l12651.htm.

rate for late 2021), as compared to more than BRL 15 billion of investment in 2010, according to data of the National Treasury Secretariat.[89]

As a tool for preserving the environment, the space plays an important role, which recently became evident in the Brazilian case. On 28 February 2021, the so-called Amazonia-1 satellite has been launched from the Satish Dhawan Space Centre in India.[90] Amazonia-1 is the first Earth observation satellite totally designed, integrated, tested and operated by Brazil, which has been launched after twelve years of development in partnership of the INPE, the AEB and the Ministry of Science, Technology and Innovation.[91] Amazonia-1 is also a first step towards a greater independence of the country from images of foreign satellites.[92]

The satellite is part of the Amazon Mission, which seeks to provide images for remote sensing to monitor the environment and agriculture in the Brazilian territory,[93] especially in the Amazon region, providing data for the INPE, the scientific community and governmental bodies, as well as integrating the Amazon Real-Time Deforestation Detection observation system (DETER, in the acronym in Portuguese) to support the surveillance and control activities.[94]

Also, as previously mentioned in Sect. 5.4, infrastructure is a crucial aspect for the development of the space sector in Brazil: the launch of satellite took place in the Indian territory because Brazil neither had a rocket that could put the Amazonia-1 in orbit, nor had the infrastructure necessary to launch rockets in the specifications required through the CLA.[95]

Therefore, even though the improvement of the infrastructure may lead to more projects integrally carried on in Brazilian soil, the development of space utilities using Brazilian know-how has not been prevented due to the lack of the appropriate infrastructure, and such utilities are often directed to environmental protection, such as the iconic launch of Amazonia-1. Despite the challenges posed in recent years and the reduced budget, space technology may be an important step towards protecting natural resources.

---

[89] Lu Aiko Otta, Gastos com proteção ambiental atingiram em 2021 menor nível em 12 anos, mostram dados do Tesouro, Valor Econômico, 23 June 2022, https://valor.globo.com/brasil/noticia/2022/06/23/gastos-com-protecao-ambiental-atingiram-em-2021-menor-nivel-em-12-anos-mostram-dados-do-tesouro.ghtml.

[90] Amazonia-1, primeiro satélite 100% brasileiro, é lançado na Índia, Institute for Applied Economic Research, 5 March 2021, http://desafios2.ipea.gov.br/cts/pt/central-de-conteudo/noticias/noticias/239-amazonia-1-primeiro-satelite-100-brasileiro-e-lancado-na-india.

[91] Ibid.

[92] Ibid.

[93] Available at: www.inpe.br/amazonia1/.

[94] Otta, Gastos com proteção ambiental atingiram em 2021 menor nível em 12 anos, mostram dados do Tesouro.

[95] Ibid.

5.5 Debureaucratization

According to the Doing Business ranking of the World Bank, Brazil is currently in the 124th position worldwide, among 190 countries ranked.[96] This ranking reflects some important characteristics that are faced by companies entering and operating in Brazil, such as the ease to start a business, register a property, get credit, pay taxes, and enforce contracts: while Brazil's best performance is identified in the enforcement of contracts (58th position), the country faces many challenges in terms of its tax system (184th position).[97]

In terms of the perception of the Brazilian population on the bureaucracy of the country, a poll carried on by the Federation of Industries of the State of São Paulo (FIESP, in the acronym in Portuguese) indicated that 84% of the population considers Brazil bureaucratic, and 78% considers that such barriers hinder development.[98]

In order to address some issues concerning the bureaucratic nature of activities developed in Brazil, the President of the Republic issued a provisional measure in 2019 creating the Declaration of Economic Freedom Rights, later converted into Federal Law No. 13.874, of 20 September 2019.[99] Provisional measures are acts enacted by the President of the Republic in urgent and relevant matters, with immediate effects of Federal Law, being submitted in sequence to the consideration of the National Congress.[100]

Federal Law No. 13.874, of 20 September 2019, established the obligation to carry out a regulatory impact assessment in the federal public administration when creating or changing normative acts, in order to assess the potential effects of the norm regarding the reasonability of its economic impact.[101] According to the OECD, the regulatory impact assessment "can underpin the capacity of governments to ensure that regulations are efficient and effective in a changing and complex world".[102]

Likewise, the Presidential Decree No. 10.139, of 28 November 2019, provided some criteria for the consolidation of normative acts hierarchically inferior to presidential decrees, such as ordinances and resolutions, also establishing the obligation

---

[96] The World Bank—Doing Business Archive, Ease of Doing Business rankings, https://archive.doingbusiness.org/en/rankings.

[97] Ibid.

[98] Flávia Albuquerque, Pesquisa mostra que 84% da população consideram o Brasil um país burocrático, Agência Brasil, 17 October 2017, https://agenciabrasil.ebc.com.br/economia/noticia/2017-10/pesquisa-mostra-que-84-da-populacao-considera-o-brasil-um-pais-burocratico.

[99] Federal Law No. 13.874, 20 September 2019, www.planalto.gov.br/ccivil_03/_ato2019-2022/2019/lei/L13874.htm.

[100] Brazilian Federal Constitution, 5 October 1988, www.planalto.gov.br/ccivil_03/constituicao/constituicao.htm.

[101] Federal Law No. 13.874, 20 September 2019, www.planalto.gov.br/ccivil_03/_ato2019-2022/2019/lei/L13874.htm.

[102] Organization for Economic Co-operation and Development, Regulatory impact assessment, www.oecd.org/regreform/regulatory-policy/ria.htm.

to revoke norms whose effects have already been exhausted or that, although still valid, its necessity or meaning could not be identified.[103]

Both above-mentioned initiatives seek to remove the complexity of the Brazilian legal system, either removing norms no longer needed or assessing the impacts created by new norms prior to their enactment. Besides these general provisions on the production and review of legal norms, a recent initiative in the space sector may be identified as another trend for simplifying the Brazilian legal system and, therefore, reducing its bureaucratic nature.

As previously mentioned, the space legal framework in Brazil includes ordinances and resolutions, such as Ordinance No. 698, of 31 August 2021, which established the REB Parts 01 and 02 containing rules regarding procedures related to the operator's license and the launch authorization.[104] Besides the importance of the subject, which has been already explored in detail by Ian Grosner, Adriana Simões and Marina Stephanie Ramos Huidobro in a work published on part 3 of this book series,[105] there is a specific initiative which shall be highlighted in this chapter.

One of the main features of the recently published REB was to "bring Brazilian regulation in line with international standards, especially the relevant norms issued by the Federal Aviation Administration (FAA) of the United States of America".[106] Specifically regarding the launch authorization regulation contained in REB Part 02, in its item 8.1.1,[107] it is established that "a party in compliance with the FAA 14 CFR Part 450 (namely the Launch and Reentry Licensing Requirements) is also complying with the new Brazilian regulation on the matter, except for some reserved items".[108]

In terms of the telecommunication regulations on exploitation of satellites, the above-mentioned Resolution No. 748, of 22 October 2021, approving the General Regulation of Satellites Exploitation, establishes that ANATEL may require compliance with provisions of the Radio Regulations enacted by the International Telecommunication Union (ITU) and guidelines for the use of outer space enacted by the United Nations Office for Outer Space Affairs (UNOOSA).[109]

By establishing some sort of automatic recognition of parties compliant with the regulation of one of the most renowned regulatory bodies in the industry,

---

[103] Presidential Decree No. 10.139, 28 November 2019, www.planalto.gov.br/ccivil_03/_ato2019-2022/2019/decreto/D10139.htm.

[104] Ordinance No. 698, 31 August 2021, www.in.gov.br/web/dou/-/portaria-n-698-de-31-de-agosto-de-2021-341897559.

[105] Grosner, Simões and Huidobro, What is Brazil Doing to Develop Its Commercial Space Program?, in: A. Froehlich, Space Fostering Latin American Societies, part 3, Southern Space Studies, Springer Nature, 2022.

[106] Ibid., p. 15.

[107] REB Part 02 of 2021, p. 9.

[108] Grosner, Simões and Huidobro, What is Brazil Doing to Develop Its Commercial Space Program?, p. 16.

[109] ANATEL Resolution No. 748, 22 October 2021, https://informacoes.anatel.gov.br/legislacao/resolucoes/2021/1595-resolucao-748.

or by establishing a potential mechanism to require compliance with international standards, the Brazilian regulation may strengthen legal certainty and reduce the transaction costs that would be incurred by a party seeking to operate space launches from Brazilian territory or to exploit Brazilian or foreign satellites, which makes Brazil even more attractive to these activities.

Therefore, adopting legal norms in line with international best practices, as recently introduced in the regulation for the space sector on launch authorization, may be identified as another way of reducing bureaucracy and making the investor's experience easier, without waiving strict safety requirements.

6 Final Remarks

As a developing country, marked by several challenges in inducing better levels of development, Brazil must act on different fronts to address its gaps. Even though law is not the only tool for addressing important issues in the Brazilian society, one may not deny the penetration that the legal system has in the various economic, social and political topics.

In order to better assess the challenges to development, the Law and Development field seems to be a suitable instrument, since it privileges the role of the legal system without underestimating the role of other sciences to build tailormade solutions, as well as understands that not necessarily the one-size-fits-all solutions are the most appropriate, depending on the country's particularities.

In this regard, it is crucial to seek multifaceted solutions to complex problems and to think outside the box, according to popular slang. The aim of this chapter was to bring another perspective for the discussions on development: besides the general and common-sense propositions usually implemented by governments, the outer space and its evolving features may be an important ally in the search for greater levels of development.

This chapter will have accomplished its purpose insofar as it has been able to demonstrate the particular challenges faced by Brazilian society, considering recognized indexes and data, and how the legal framework related to the space may contribute to overcome, at least in part, such challenges. Despite the current effort to analyze the development of societies through space under the lens of Brazil and the law, contributions on the role of space in the development of other nations, especially in Latin America, and the inputs of other areas of knowledge on the topic, are always welcome.

Adriana Simões is a partner at Mattos Filho, based in the city of São Paulo, Brazil. She specializes in aviation law, contracts and infrastructure. She is a member of the Aviation Law Committee and the Space Law Committee of the International Bar Association (IBA), the International Aviation Womens Association (IAWA), Latin American and Caribbean Air Transport Association (ALTA), Contact Group of the Aviation Working Group (AWG), the Brazilian General Aviation Association (ABAG), as well as the Brazilian Bar Association Aeronautical Law Commission, São Paulo Section (OAB/SP).

Felipe Kotait Buchatsky is a lawyer at Mattos Filho, based in the city of São Paulo, Brazil. He is an Academic Master Student in Law and Development at the São Paulo Law School of Fundação Getulio Vargas (FGV DIREITO SP) and a Bachelor of Laws by the same institution. He specializes in aviation law, contracts, constitutional law and law and development. He is enrolled to the Brazilian Bar Association, São Paulo Section (OAB/SP).

Honduras in Space so Far: A Central American Approach

Javier Mejuto

Abstract

This chapter intends to give a vision of the efforts in the special subject of the Morazán Project in Honduras from a Central American perspective. A small retrospective of the regional space missions is carried out to date to explain in detail the only Central American space project in development today, the Morazán Project. Finally, some of the trends in space issues in the Central American region are exposed.

1 Introduction

The arrival of the first space projects in Central America has been long awaited, however, since the launch on 11 May 2018 of the first Central American satellite,[1] the Batsú-CS1, developed by the Central American Aeronautics and Space Association and operated by the Technological Institute of Costa Rica, the interest in space and the benefits that the population can obtain from it have grown exponentially in the Central American region, to the point that we could speak of a real space race for the countries of the region.

[1] SICA note about Batsú-CS1 launch, https://www.sica.int/consulta/Noticia.aspx?Idn=112831&idm=1 [all websites cited in this publication were last accessed and verified on 18 August 2022].

J. Mejuto (✉)
Space Sciences Faculty, Archaeoastronomy and Cultural Astronomy Department, Universidad Nacional Autónoma de Honduras, Tegucigalpa, Honduras
e-mail: javier.mejuto@unah.edu.hn

Centre for Astrophysics, Institute for Advanced Engineering and Space Sciences, University of Southern Queensland, Toowoomba, Australia

A. Froehlich (ed.), *Space Fostering Latin American Societies*, Southern Space Studies,
https://doi.org/10.1007/978-3-031-20675-7_2

After this satellite launch, a product of the Irazú Project, the first Guatemalan satellite followed, and the first with technology developed in Central America, developed by the Universidad del Valle de Guatemala (UVG) on 28 April 2020.[2] Likewise, the Morazán Project,[3] which will be the first Honduran satellite and the first Central American satellite, in collaboration with universities from three Central American countries: the National Autonomous University of Honduras, the University of Costa Rica, and the San Carlos University of Guatemala. These are the first steps of the Central American growing space ecosystem.

2 Morazán Project

The Morazán Project is an entirely academic project led by the Faculty of Engineering, in conjunction with the Faculty of Space Sciences, both of the National Autonomous University of Honduras. It is an interdisciplinary and international effort, as academics and students from three countries of the Central American region are represented under the grouping of the Central American Integration System (SICA), thereby seeking integration in space. In addition to these stakeholders, the project is supported by Japan Aerospace Exploration Agency (JAXA), the Kyushu Institute of Technologies in Japan (Kyutech) which will be key in the technical development of the satellite, the Federated College of Engineers and Architects of Costa Rica (CFIA) as well as the Mauritius Research and Innovation Council (MRC) of the Republic of Mauritius.

The Central American region is considered a multi-threat zone due to its geographic location, in terms of potential hazardous impact from hurricanes, high seismicity and volcanism, loss of biodiversity, and other hydrometeorological activities. This set and combination of threats for Central America "is considered the second most vulnerable to weather risks".[4] In addition, Central America, due to its economic, political and, social conditions, is considered as a region with a level of vulnerability that determines the impact of disasters.

In 2010, after the commitment made by the region by adopting the Sendai Framework for Disaster Risk Reduction 2015–2030,[5] the Central American Policy for Comprehensive Disaster Risk Management (PCGIR) conducted a first SICA Summit Meeting. The PCGIR has as main objectives:

(a) Disaster risk reduction in public and private investment for sustainable economic development,
(b) Development and social compensation to reduce vulnerability,

[2] QUETZAL-1 Satellite website, https://www.uvg.edu.gt/cubesat/.

[3] MORAZÁN PROJECT website, http://proyectomorazan.space.

[4] CEPREDENAC–SICA (2017). PCGIR–MSRRD 2015–2030, https://ceccsica.info/sites/default/files/docs/Politica%20Centroamericana%20de%20Gestion%20Integral%20de%20Riesgo.pdf.

[5] Resolution 69/283 of the General Assembly of the United Nations, https://www.un.org/en/development/desa/population/migration/generalassembly/docs/globalcompact/A_RES_69_283.pdf.

(c) Disaster risk management and its relationship with change climate,
(d) Territorial management, governability and, governance and
(e) Management of disasters and recovery.

Thus, the Sendai Framework recognizes that the application of Geospatial Technologies at different scales contributes both to the monitoring of the relevant United Nations SDGs and to a responsable disaster risk management; considered valuable tools due to their wide applicability and their contribution to the understanding of geosystems.

The high exposure to hydrometeorological extreme events over the last three decades—more than twenty occurred yearly in Central America—became this type of natural event with the greatest impact on the Central American population, both in terms of personal and economic casualties, endangering food and water sovereignty. An example of this could be seen in the impact of two major hurricanes in the span of 13 days, Hurricane Eta on 3 November 2020, and Hurricane Iota on 16 November 2020. These climatic events reached Categories 4 and 5 respectively on the Saffir-Simpson scale[6] (Fig. 1). The Pan-American Health Organization (OPS) estimates in almost ten million affected people after the passage of these two extreme weather events. Other sources show devastating numbers in fruit and vegetable losses as a consequence of the flood in a huge part of Honduras, Nicaragua, and Guatemala for several weeks. According to data from the United Nations Economic Commission for Latin America and the Caribbean (CELAC) report, these events produced a reduction of 0,8% in GDP growth for the year 2020.[7] Central America will face these types of events harder and more frequently as a result of climate change in the coming years and, therefore, it must take steps towards resilience.

2.1 Scientific Mission

Given the reality of natural risks in Central America, the scientific mission of the Project was oriented to have an impact on the reduction of flood disasters. The Project takes one hydrographic basin for each of the three countries involved as the unit of analysis, being defined as the area bounded by watersheds, where precipitation forms the main course of a river. The first watershed is that of the Samalá River in Guatemala, with an area of 1.633 km^2 and an average rainfall of 81,91 mm per year. The next is the Ulúa River basinwatershed in Honduras, with an area of 22.817 km^2 and an average annual rainfall of 87,63 mm, and finally

[6] NASA and Historical Hurricanes Eta and Lota: https://www.nasa.gov/feature/goddard/2021/after-historic-hurricanes-eta-and-iota-nasa-helps-prep-central-america-for-disasters-to-come.
[7] UNITED NATIONS, CEPAL, Evaluación de los efectos e impactos causados por la tormenta tropical Eta y el huracán Iota en Honduras, Signatura: LC/TS.2021/22236. CEPAL, BID, May 2021, https://www.cepal.org/es/publicaciones/46853-evaluacion-efectos-impactos-causados-la-tormenta-tropical-eta-huracan-iota.

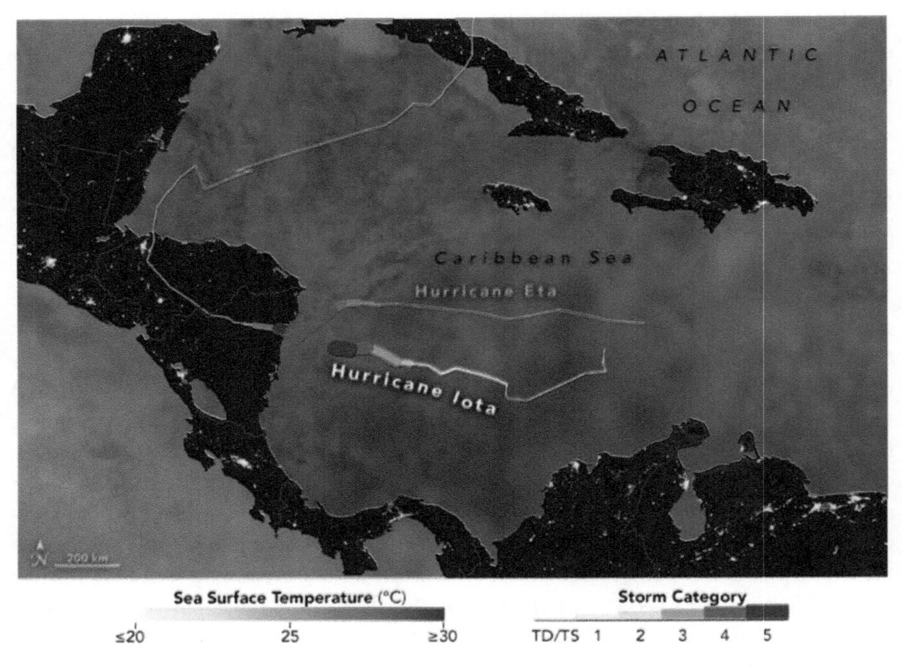

Fig. 1 Hurricane Eta and Ioeta in 2020—Path and Categories with sea surface temperatures in the Central American Region[8]

the Matina River basinwatershed in Costa Rica, with an area of 1.428 km^2 and an average annual rainfall of 218,08 mm.

The Morazán Project's scientific team pursues the combination of in situ climate data with freely accessible information from Earth observation satellites, to develop a near early warning system for natural events such as floods or landslides. Likewise, a real-time data collection system based on simple and accessible infrastructure such as the APRS Protocol or other protocols based on low-cost analog infrastructure will be carried out, especially useful in regions with little or no availability or access to communication infrastructure. The Project is a proof of concept that aims to demonstrate the future value of a constellation of nanosatellites for packet radio communication to support the monitoring and early warning of extreme natural events. Furthermore it will test the satellite-based VHF and UHF radio frequencies capacity for communication in post-event emergencies in areas of the world where access to communication infrastructure is poor or vulnerable and thus to be mitigated.

[8] NASA's Earth Observatory/Using GOES 16 imagery courtesy of NOAA and the National Environmental Satellite, Data, and Information Service (NESDIS), data from the Multiscale Ultrahigh Resolution (MUR) project, and Black Marble data from NASA/GSFC.

Fig. 2 Placement of data collecting monitoring stations prototypes for the Morazán Project in the field

To achieve these Morazán Project goals a network of data collecting monitoring stations will be placed in these watershed (Fig. 2) with three possible scenarios: Pre-event, event and post-event. In the pre-event scenario, the stations will record relevant hydrometeorological variables (water level, humidity and soil saturation, river flow, relative humidity, wind speed, etc.). These variable data are transmitted to the satellite, which in turn will be received by the Project's ground station. There they will be stored and analyzed for the performance of behavior models of the respective watersheds under study. They will also be used as proof of concept for future missions and a possible future satellite constellation that implements the first Central American early warning system based on satellite technology.

In a possible flood event, such as the arrival of a hurricane, the data will be prioritized based on the storm trajectory and the basin watershed that will be most affected. These data, once uploaded to the satellite, will be evaluated and analyzed at the ground station to generate vital information. As a result corresponding necessary alerts could be issued both to the local population in the watershed area, to the general population and to emergency response services. These same information could also be received in the communities that so wish for community risk management and independent decision-making.

In the event that a disaster has occurred with isolated populations without communication, the post-event scenario will be activated. In this case, the satellite could receive help and distress signals from the communities with information that will allow the members of the emergency services to make an efficient intervention on the ground.

2.2 Ground Segment and Educational Mission

The Morazán Project satellite will follow the CubeSat 1U standard. Design, assembly, and testing will be done through a collaborative effort in which teams from the main public universities of the Central American countries are represented in the project. The management of the satellite, once in operation, will also be carried out collaboratively with different teams in Central America through ground stations located on university campuses and radio amateur organizations.

The ground segment is the final component of the Morazán Project. Its functions are tracking the satellite, monitoring its status, uploading commands, and most importantly downloading the data that originated from the hydrographic basins. The ground station is being developed jointly with Honduran Association for Amateur Radio, using recommendations from M2 Antenna Systems Inc., which offers solutions for amateur radio satellite ground stations. The main objective was to create an affordable low-cost station for operations in the UHF band. This can be accomplished by using COTS (Commercial off-the-Shelf) components. The ground station component relates to the actual hardware required to contact the spacecraft, while mission control focuses more on the software needed to operate the mission. The final element of the ground segment is used to turn the raw data acquired by the ground sensors to produce scientific imagery to support decision makers and that will be attractive to the public and raise interest in the mission. This infrastructure works for satellite construction are to be devised and developed in a scientific academic environment. The reason for this design lies in the fact that one of the Project objectives is to be the initiator of future space projects and programs for which a critical mass of professionals is needed. The universities are the places, together with the different educational levels in the country, in which we look for these scientific vocations among the students who will be able to participate both in the development and construction of the satellite and in the operation management phase. Thus, an academic ground station will allow students and different other areas of professionals to use it as a learning and education tool to acquire competencies in STEM for several areas and subjects in their studies.

Following the same idea, the clean room necessary for the assembly of the satellite itself has been designed. This will not only support space technology but can also be used for different disciplines, especially for research and development of medicines, in close collaboration with the Faculty of Chemistry and Pharmacy of the National Autonomous University of Honduras. As can be seen, we have tried to ensure that all investments in infrastructure are of the greatest interest and widest use within the universities and for the benefit of the largest possible population.

Another misión goal of the Project is to promote Satellite Technology and Science in schools, colleges, and universities, through the management of weather stations that communicate via satellite using the APRS Protocol. This aspect is one of those that we consider most important due to the forward-looking creation of schooling data for the Central American Region in general and Honduras in particular. Actual data provided by UNICEF show that 44% of boys and girls between 3 and 17 years old are out of the educative system.[9] Boys and girls in the groups with the highest level of exclusion from educational coverage are those who live in rural areas (48,7%), those between 3 and 5 years old (65%), adolescents between 12 and 14 years old (55,1%) and between 15 and 17 years old (74,6%). The main cause of non-attendance to the third cycle of basic education is the lack of economic resources (46%). In total, 4% left school to go to work and 28% do not want to continue studying because they do not perceive that education opens the doors to the world of work. On the other hand, in the third cycle of basic education, 28,5% of teenagers are at moderate risk of dropping out of school and another 28% are at serious risk. Regarding the quality of learning, only 55% of elementary school students performed satisfactorily or advanced in mathematics and only 63% in reading on national assessments. Another of the main causes of this problem is the lack of bilingual or intercultural education putting the focus on the cultural diversity of the country and the different languages and worldviews that exist.

In this context, we think that using this Morazán Project can, in the long run, motivate students from the lowest educational levels in scientific subjects. Thus it will allow them to maintain their interest in continuing to study and open opportunities in the future for young Central Americans both in the space field and in other natural scientific fields. To do this, the simplest possible equipment has been designed, made up of a Yagi antenna and a free available cell phone application, that young people in schools can independently receive and interpret the data received from the "own" national satellite. For this specific purpose, the Morazán Project satellite will have a camera that will produce photographs of the Region, being the first time that Central Americans can receive images of their local area through a device made in their region.

2.3 Space and Indigenous Peoples

According to the United Nations Permanent Forum on Indigenous Issues, the indigenous population is estimated at approximately 370 million people living in more than 90 countries.[10] This constitutes 5% of the world population but also represents 15% of the extremely poor population and 33% of the rural poor population. In fact, not only are there more indigenous people in a situation of poverty, but their level of poverty is comparatively much higher. The lands they inhabit constitute 80% of the planet's biodiversity, including access to a quarter of the drinking water and land. This is also a reality that is lived in Central America

[9] UNICEF, Inclusive and quality education for every girl and boy, https://www.unicef.org/honduras/que-hacemos/educación-de-calidad.

[10] United Nations Permanent Forum on Indigenous Issues, https://www.un.org/esa/socdev/unpfii/documents/5session_factsheet1.pdf.

where its cultural aspects are even more relevant to the water-related worldviews that form a single system along with agriculture, climate, and celestial space. A paradigmatic case is the case of the Lenca culture in Honduras. This Lenca culture inhabits one of the watersheds under study by the project, the Ulúa River watershed in Honduras. A total of five indigenous cultures share this watershed, a similar cultural context can be seen in the Samalá watershed in Guatemala and Matina watershed in Costa Rica.

The relationship between indigenous peoples and the sky in Central and Mesoamerica has been extended for more than 3.000 years forming their legacy as astronomical heritage (Fig. 3). We can see evidence relating to the practice of astronomy and to social uses and representations of astronomy in all archaeological periods in this cultural region. It exists in the form of the tangible remains of monuments, archaeological sites, and landscapes with a link to the skies that constitute a well defined physical property. It can also involve movable objects such as instruments and archives, intangible knowledge—including indigenous knowledge still preserved in the world today—and natural environments. These support human interest in astronomy, for example through the cultural use of their horizons or dark night skies.

Indeed, we can find this type of heritage throughout the entire Central American Region in archaeological sites such as Uaxactún in Guatemala, Copán in Honduras, and Cihuatán in El Salvador are witnesses of an ancient astronomical heritage, but also refers to a broad set of material evidence that served as keys to the interpretation of celestial bodies, with social uses and representations.

But this heritage lies not only in the past but in the present with contemporary indigenous peoples who have been historically excluded from the national infrastructure and social welfare programs for ethnic and socio-economic reasons. The social exclusion relegates a significant part of the Central American population to a situation of extreme vulnerability to all kinds of risks, especially natural hazards such as those produced by seismic and volcanic events and of course, floods.

On the other hand, these populations constitute the first line in the fight against climate change, food security and access to drinking water. Undoubtedly, resilience to these climate change processes involves strengthening the empowerment of these communities based on self-management of risk and vulnerability, combining ancestral knowledge with new technologies at our disposal.

Understanding and valuing the long tradition and relationship with the celestial space of the native Central American peoples, we see this Morazán Project as the next step. However, the Project does not intend to carry out work for the indigenous peoples, but rather with them. That is why joint workshops are being held in the communities that are distributed by the watershed under study with the following objectives:

(a) Socialization of the project,
(b) Understanding the indigenous worldviews,
(c) Inclusion of members of indigenous communities in the project work team,

Fig. 3 Maya astronomer from the Madrid Codex (p. 34)

(d) Understanding of the accumulated ancestral knowledge for its use in the identification of flood zones, and
(e) Organizing training workshops in the use of the data that the project will provide.

Including indigenous peoples in the Morazán Project from the beginning, it is expected that the positive impact on the communities in the study area will be much greater than that which would be achieved with a classic non-inclusive approach. Traditionally one of the products that would be obtained from this type of project is a national risk management plan, but in this case as a first step only a local community-based risk management plan. In the same way, a water management plan will not be obtained, but community-based water security plan. It is intended that Morazán Project is not an intervention project but rather a self-management project that promotes the empowerment of indigenous communities, being able to make decisions based on space-assisted data for the benefit of their community and people's lives.

3 Space Ecosystem and Challenges

The moment that Central America is going through in space subjects is tremendously exciting, but it is very incipient. In most of the countries of the Region, investments in Science is below 1% of GDP[11,12] and there is no critical mass for academics and technicians in the Region. However, the space sector can be, and in a certain way already is, a motor for the development of science in the Region. In recent years we have been able to see progress in the development of space public policy in different Central American countries and even the opening of national and regional space agencies with those responsible for leading national efforts in this field. Undoubtedly, much remains to be done, but the progress that has been seen in just over four years since the launch of the first Costa Rican satellite has not been achieved in other fields for decades. Some examples of the progress that is expected in the coming years are regional policies that allow us to expect that this growth of the sector will continue, that regional initiatives that will promote joint projects and programs, and therefore, greater resources for their realization. I would like to mention here two initiatives, in particular, the Central American Astronautical Network (RAC),[13] a network of students and young professionals dedicated to creating ties that allow the development of space sciences in the Region. On the other hand, the Latin American and Caribbean Space Agency (ALCE) called to be the catalyst for Latin American space programs, becoming in the near future one of the weighty actors on the world stage.

Without the intention of being complete, a series of problems are proposed below that must be solved in the short term so that this development of the local space sector continues:

(a) Create an efficient national and regional customs procedure for the temporary or permanent import of scientific equipment.
(b) Develop national and regional space public policies.

---

[11] BID 2010, Science, technology and innovation in Latin America and the Caribbean: A statistical compendium of indicators, https://publications.iadb.org/publications/english/document/Science-Technology-and-Innovation-in-Latin-America-and-the-Caribbean-A-Statistical-Compendium-of-Indicators.pdf.

[12] RICYT 2021, El estado de la Ciencia, Principales Indicadores de Ciencia y Tecnología Iberoamericanos/Interamericanos, http://www.ricyt.org/wp-content/uploads/2021/11/El-Estado-de-la-Ciencia-2021.pdf.

[13] RAC web: https://redespacial.com/.

(c) Convert the space sector into a priority of the Central American countries as an element of development, betting firmly on projects and programs that have a positive impact on the population. This must occur in two aspects; a marked and sustained political will towards science and sufficient economic resources.
(d) Signature and/or ratification of International Treaties Space Law Treaties and Principles.
(e) Create a legislative environment that facilitates the creation of a spatial ecosystem in all its branches.

As it can be seen, the future in space is promising but we cannot fix our eyes on space without having our feet on the ground. In the effort of making projects we cannot forget for whom we carry out these projects, we must focus on inclusive programs, honoring tradition and ancestral heritage that solve the problems of the region, thus improving the quality of life of our citizens.

***

Javier Mejuto is Professor at the Space Sciences Faculty at the National Autonomous University of Honduras, where he holds the position of Head of the Department of Archaeoastronomy and Cultural Astronomy and Adjunct Professor at the University of Southern Queensland in Australia. He is the Honduran Science Coordinator of the Morazan Project and President of the IAU National Astronomy Commission (Honduras) of the International Astronomical Union (IAU). He also serves as Co-Chair of the Archaeoastronomy and Astronomy in Culture Working Group of the IAU and part of the Ethnoastronomy and Intangible Astronomical Heritage and the Astronomical Heritage in Danger IAU Working Groups. He is expert member of the ICOMOS International Committee on Intangible Cultural Heritage (ICICH) and he leads the Central American and Caribbean Initiative for Astronomical Heritage. Between other proffesional societies he serves as Secretary of the Inter-American Society of Astronomy in Culture, member of the International Society of Archaeoastronomy and Astronomy in Culture, fellow of the European Society of Astronomy in Culture and member of the Committee for the Cultural Utilisation of Space (ITACCUS) and Latin America and Caribbean Regional Group (GRULAC) of the International Astronautical Federation (IAF). His research and work lines cover the inclusion of Developing Nations and underrepresented groups, especially indigenous peoples, in Space exploration and the access of Spatial technology.

Democracy Through Connectivity: How Satellite Telecommunication Can Bridge the Digital Divide in Latin America

Lauryn Lee Hallet and Marieta Valdivia Lefort

Abstract

Space telecommunication offers many benefits. It can be used to access news sources and share real-time information from virtually anywhere on the planet. It can also empower citizens with open-source informal education tools. This article argues that space technologies can support democracy in Latin America by reducing digital inequalities. To do this, we briefly introduce the notion of democracy and its current state in Latin American, followed by an explanation on how the space sector can help reduce inequalities and foster democracy by, for instance, broadening access to telecommunication services in remote and rural areas. We then shed light on the democratisation of space technologies in Latin America; the evolution of the access to the space sector itself and the access to space telecommunication services. Finally, we present a set of recommendations for future actions at both regional and national levels, including increasing the coverage zones of telecommunication satellites, improving latency in the region, securing people's access to electricity and the necessary local ground infrastructure, and encouraging both local private investment and regional cooperation.

L. L. Hallet (✉)
Centre for Youth and International Studies (CYIS), Brussels, Belgium
e-mail: lauryn.hallet@cyis.org

M. V. Lefort
Institute of the Americas, University College London (UCL), London, UK
e-mail: marieta.constanza@gmail.com

A. Froehlich (ed.), *Space Fostering Latin American Societies*, Southern Space Studies,
https://doi.org/10.1007/978-3-031-20675-7_3

1 Introduction

The state of democracy in the Latin American region has shown significant tensions in recent years. According to International IDEA's regional report, the state of democracy in the region has deteriorated; certain countries still harbour authoritarian regimes or hybrid ones oscillating between authoritarianism and some form of democracy.[1] In addition, the study found that the Covid-19 situation has impacted democracy in the region, including "freedom of expression, personal security, and integrity".[2]

In this scenario, discussions on palliative measures have been common at both political and social levels, with one of these being the reduction of inequality and the democratisation of access to services to improve people's quality of life. In this sense, the significant difference between the rural and urban sectors in terms of access to services has also been widely discussed, with the inhabitants of rural areas being the protagonists in demanding improvements in their quality of life through, among other aspects, improvements in connectivity. Thus, the space sector and satellite development present an opportunity for improved connectivity, requiring the creation of policies and measures that contribute to this.

Therefore, this document discusses how space technologies can support the improvement of democracy in Latin America by reducing urban–rural inequality in issues such as connectivity through telecommunication. Thus, the Sect. 1 presents a brief discussion of the concept of democracy at the political level, and its status in the Latin American context. The Sect. 2 explains how the space sector, and in particular telecommunication, can help both reduce inequality and foster democracy in the region by broadening access to connectivity services in remote and/or rural areas. The Sect. 3 then explores the democratisation of space technologies in the region by studying both the evolution of the access to the space sector itself and the access to space telecommunication services. Finally, Sect. 4 lays a set of takeaways and recommendations for future actions, indicating measures that could be taken to improve people's quality of life through the use of satellite technology.

2 Democracy in Latin America

Agreeing on a single and universal definition of democracy has not been possible. In the current literature, different definitions for this concept can be found at the theoretical level, as well as different ways and scales for measuring its presence and quality in different countries worldwide. This diversity has hindered consensus on a universal definition, although there is agreement on certain elements that are fundamental to talk about democracy and a democratic regime.

[1] International IDEA, "The Americas: Democracy in Times of Crisis3, 2021, www.idea.int/gsod/las-americas-eng-0 (all websites cited in this publication were last accessed and verified on 29 August 2022), International IDEA, "The Americas", 2021, www.idea.int/gsod/las-americas-eng.
[2] Ibid.

In this respect, Spicker argues that democracy is indeed a multifaceted concept with a set of meanings and interpretations, such as "a system of government, defined by a set of institutional arrangements" and prescriptions for governance.[3] Similarly, Storm notes that the concept of democracy is used to describe different circumstances, such as a situation where, among other aspects, basic civil liberties and rights are respected and also protected.[4] Furthermore, and along the same lines, Santos González and Martínez-Martínez point out that one of the levels at which democracy develops is at the political level, and it does so as a form of organisation of society that includes decision-making in favour of the majorities and the collective good.[5]

In the Latin American context, durable democracies were progressively born during the 1980s and 1990s, but they have not been free from lingering flaws and barriers, such as "persistent and pervasive inequality, predatory relations between rulers and ruled, and clientelism".[6] Moreover, Feldmann, Merke and Stuenkelthat add that regional efforts to preserve democratic order in some countries have been moderately effective,[7] while others have experienced situations that directly threaten the democratic political regime such as "repeated presidential or congressional crises (...) [and states that] do not enforce laws effectively and evenly throughout the country's territory".[8] In this respect, Peeler argues that, among other aspects, liberal democracies in the region are mostly of very low quality, and "they have largely failed to deal effectively with the region's massive social problems, such as the poverty endured by the majority of the population (...) to reverse the growing, increasingly visible inequality of Latin American societies (...) [and] to build states strong enough to actually carry out a coherent policy agenda".[9]

Recent studies on the quality of democracy in Latin America, such as the Congressional Research Service report (July 2022),[10] suggest that the scenario at the

---

[3] Paul Spicker (2008), Government for the people: The substantive elements of democracy, *International Journal of Social Welfare*, *17*(3), p. 251, https://doi.org/10.1111/j.1468-2397.2008.00556.x.

[4] Lise Storm (2008), An elemental definition of democracy and its advantages for comparing political regime types. *Democratization*, *15*(2), 215–229, https://doi.org/10.1080/13510340701846301.

[5] Yissel Santos González and Oscar Martínez-Martínez (2020), La insatisfacción con la democracia en América Latina. Análisis de factores económicos y políticos en 2017, *Universitas*, 32, 157–174, https://doi.org/10.17163/uni.n32.2020.08.

[6] John Peeler (2022), *Building democracy in Latin America*, Lynne Rienner Publishers, p. 32.

[7] Andreas E. Feldmann, Federico Merke, and Oliver Stuenkel (2019), Argentina, Brazil and Chile and democracy defence in Latin America: Principled calculation, *International Affairs*, *95*(2), 447–467, https://doi.org/10.1093/ia/iiz025.

[8] Sebastián L. Mazzuca and Gerardo L. Munck (2020), *A middle-quality institutional trap: Democracy and state capacity in Latin America*, Cambridge University Press, p. 1.

[9] Peeler (2022), *Building democracy in Latin America*, p. 208.

[10] US Congressional Research Service report *Democracy in Latin America and the Caribbean: A Compilation of Selected Indices*, 11 July 2022, https://crsreports.congress.gov/product/pdf/R/R46016.

regional level is not very optimistic. Particularly, this report compiles results from different prestigious indexes on the quality of democracy in the region, such as the *Bertelsmann Transformation Index* (BTI), the Economist Intelligence Unit (EIU)'s *Democracy Index*, and the Varieties of Democracy Institute (V-Dem)'s *Liberal Democracy Index*. In detail, BTI 2022 points out that, in general, there is a continuous decline in the quality of democracy in the region, which is reflected, among other aspects, in the political turmoil observed in the majority of the Latin American countries. The report also highlights that some countries in the Latin American and Caribbean regions, such as Chile, Costa Rica, Uruguay and Jamaica, have been able to maintain their level of democracy despite myriad transformation challenges. Similarly, although with some discrepancies, the EIU's Democracy Index 2021 suggests that the region's overall score significantly declined from 2020 to 2021, highlighting five major downgrades: Chile, Ecuador, Mexico, Paraguay, and Haiti. In the same line, the V-Dem 2021 report notes that significant regressions on democratisation can be observed in twice as many countries, highlighting Brazil, El Salvador, Nicaragua, and Venezuela.[11]

3 Space and Connectivity for Democracy

Considering the above, discussions on policies that correct the status and the perception of democracy are constantly present at the political level and strongly demanded by citizens throughout the region. In this regard, decentralising policy-making and designing policies that are applicable and effective throughout the national territory can help tackle some aspects of inequality and improve citizens' lives by, for instance, democratising access to services in rural and/or remote areas.

Issues such as the improvement of technologies for connectivity and internet access are linked to the improvement of democracy in the region, with the space sector being a valuable opportunity to achieve this. Currently, the internet is an essential service for people worldwide. This is one of the elements that represents and reinforces the phenomenon of globalisation, and as such, the creation of services, content, and new cultural and social practices associated with it have made it indispensable in people's daily lives.[12] The internet has not only allowed more people to be connected all around the world, but has also allowed the democratisation of access to information and services from everywhere, including rural and/or remote locations. Nevertheless, internet access in certain areas of the region is

[11] US Congressional Research Service report *Democracy in Latin America and the Caribbean: A Compilation of Selected Indices*, 11 July 2022, https://crsreports.congress.gov/product/pdf/R/R46016.

[12] Eduardo Villanueva-Mansilla, 2020, ICT policies in Latin America: Long-term inequalities and the role of globalized policy-making, *First Monday*, *25*(7), https://doi.org/10.5210/fm.v25i7.10865.

still an unresolved issue, for which satellite-based internet technology presents an alternative broadband solution as this can be reached from every part of the world, even in places where electricity is not a constant commodity.[13] Moreover, given the geographic characteristics of the Latin American region (e.g. thousands of kilometres of mountains and highlands, and hundreds of volcanoes),[14] satellite telecommunication could better respond to the specific requirements and demands of people facing connectivity precarity due to remoteness from large cities and/or more populated places, or even poverty. Therefore, this satellite-based technology might contribute to tackling the effects of inequality in this respect.

Broadly, telecommunication systems can be put into different categories, such as wired and wireless, fixed and mobile, optical and radio-frequency, or terrestrial-based and satellite-based. Simply explained, for satellite communication, a transmitting ground station sends a signal to the satellites (uplink) that is then sent back to another ground station, the receiver (downlink). Sometimes, the same station can both transmit and receive signals (e.g. VSAT, or very-small-aperture terminal, which are small parabolic antenna dishes). The communication is therefore travelling from one point to another on earth through a relay in space, meaning that satellite communication also requires hardware on the ground. For terrestrial communication, there is an exchange of signals between antennas on the ground. In this case, the communication travels between two points on earth's surface. A relay or repeater can be placed to extend the reach but this relay will be an installation on earth. A type of terrestrial communication transmission worth mentioning is fibre optics, where signals are transmitted through physical cables in the form of light (where satellite communication transmits signals through electromagnetic waves).

Each type of communication presents both benefits and challenges. This is why, generally, a combination is desirable to ensure efficiency and continuity of the services. On the one hand, satellite communication provides a wider area of coverage, is more precise, and allows for better quality transmission, while relying on limited and/or relatively rapidly decaying resources (e.g. the allocated bands according to International Telecommunication Union's-ITU's-spectrum management processes and the satellite itself, which can have a life expectancy of under 5 years for small

[13] Anjali Yadav, Manthan Agarwal, Somya Agarwal, and Sachin Verma (2022), Internet From Space Anywhere and Anytime-Starlink, *SSRN*, https://papers.ssrn.com/sol3/Delivery.cfm/SSRN_ID4160260_code5284765.pdf?abstractid=4160260&mirid=1.

[14] Diane Bourdeau et al., "South America: Physical Geography", *National Geography*, 29 July 2022, https://education.nationalgeographic.org/resource/south-america-physical-geography, WMO, "WMO issues report State of Climate in Latin America and Caribbean", 22 July 2022, https://public.wmo.int/en/media/press-release/wmo-issues-report-state-of-climate-latin-america-and-caribbean.

Low-Earth orbit-LEO-satellites). Signals, especially those coming from Geostationary orbit (GEO) satellites, will also experience an increased delay.[15] Satellite communication is most often used for mobile and wireless services. On the other hand, terrestrial communication covers a more restricted area and provides a lower quality transmission, but the latency is lower. Depending on the topography of a territory and other environmental conditions, signals could be obstructed; for instance, mountains could hamper the transmission and reception of signals, and both satellite and terrestrial communication can be disrupted by meteorological events.[16] Fibre optics is particularly suited for urban areas and soils that can easily be drilled, as well as for fixed applications and covering long distances. It also allows for a particularly high speed data transmission and experiences virtually no delay.

Overall, space services derived from telecommunication can be used to improve citizens' lives in several ways, including enabling people to get access to formal or informal education (particularly after the wave of digitisation brought on by the Covid-19 pandemic), to get informed about what is happening nationally and internationally through different information channels, to vote remotely, and even to allow people living in countries where crises such as dictatorships and/or armed conflicts are unfolding to find out what is really happening there through real-time personal videos or broadcasted news.

4 Democratisation of Space Technologies in Latin America

The threshold to access the space sector and its services is lower than ever before. Thanks to NewSpace and the booming of small satellites, more countries now have the capabilities to manufacture and operate their own satellites. Latin America is not failing to pick on that trend.

[15] GEO satellites (i.e. orbiting at around 35.000 km) are higher than LEO satellites (i.e. between 160 and 2.000 km). Because of that, they will cover wider areas (3–4 GEO satellites working together are enough for global coverage), but the signal will have to travel a longer distance (high latency). LEO satellites cover a more limited area (a high number of satellites working together are needed for global coverage) but are closer to the ground (lower latency). GEO satellites go slower and require more energy and LEO satellites are faster and require less energy. While satellites in GEO will traditionally have a mass of several tons, LEO satellites can be as small as a few hundreds grams.

[16] See: Muhammad Zubair et al., "Atmospheric influences on satellite communications", Przeglad Elektrotechniczny 85, no. 5 (2011), www.researchgate.net/publication/266522655_Atmospheric_influences_on_satellite_communications.

Table 1 Latin American satellites launched before 2010, and between 2010 and 2022 (Compilation of Gunter's space page, UCS satellite database, nanosats.eu)

	Argentina	Bolivia	Brazil	Chile	Colombia	Costa Rica	Ecuador
<2010	8	0	6	1	1	0	0
2010–22	40	1	18	5	1	1	4
	Guatemala	Mexico	Paraguay	Peru	Uruguay	Venezuela	Total
<2010	0	7	0	0	0	1	24
2010–22	1	8	1	5	1	2	88

4.1 Access to the Space Sector

When considering all uses and categories, i.e. telecommunication, earth observation, education, and research & development,[17] a total of 88 satellites have been launched in the Latin American region between January 2010 and July 2022, with around 20% of them being for telecommunication uses.[18] Furthermore, for the entire period predating 2010, there were 24 satellites, with 14 of them being attributed to Brazil and Argentina alone. To date, over half of Latin American satellites are from Argentina.

In this respect, Table 1 shows that the number of countries with access to space has more than doubled, going from 6 before 2010, to 13 since then. This is not merely a matter of buying and operating satellites; 9 countries in the region have their own indigenous manufacturing capabilities. While Brazil and Argentina have gained more advanced expertise with the respective construction of satellites of up to 650 kg and 3 T, Chile, Costa Rica, Ecuador, Guatemala, Mexico, Peru, and Uruguay have built small satellites below 10 kg, most of them Cubesats.

As shown in Table 2, the biggest concentration of satellites is in the small satellite range, i.e. between pico and mini, meaning that most of the countries who manufacture and/or operate satellites in the region are enabled to do so through small satellites. The rise of NewSpace and the consequent growth in the launch of this type of satellite has therefore lowered the access threshold for Latin American countries.

Returning to the previous point on indigenous capabilities, Argentina is the only country in the region with indigenous capabilities in both the manufacture and operation of telecommunication satellites. As of July 2022, there are three telecommunication systems in orbit: ARSAT 1, ARSAT 2 and MDQubeSAT1 (Table 3). The two ARSAT satellites cover Argentina and most of the Americas with telecommunication services, data transmission, internet, IP telephony, and

[17] There are no known PNT or space exploration capabilities in the region yet.

[18] As there is not one open source and exhaustive database containing all necessary information, a cross-reference of different lists had to be made, which can lead to inconsistencies. Nevertheless, these do not affect the arguments developed throughout the essay. The main sources for the authors' compilation of data were Gunter's Space Page, UCS Satellite Database, and nanosats.eu.

Table 2 Classification of Latin American satellites by mass (Compilation of Gunter's space page, UCS satellite database, nanosats.eu)

Mass category	Mass in kg	N. operated	Countries operating	N. manufactured (from n. operated)	Countries manufacturing
Large	>1.000	20	4 (ARG, BOL, BRA, MEX)	4	1 (ARG)
Medium	500–1.000	3	2 (VEN, BRA)	1	1 (BRA)
Mini	100–500	4	3 (BRA, CHL, PER)	2	1 (BRA)
Micro	10–100	39	2 (ARG, BRA)	33	2 (ARG, BRA)
Nano	1–10	31	10 (ARG, BRA, CHL, COL, CRI, ECU, GMT, MEX, PER, URY)	27	9 (ARG, BRA, CHL, CRI, ECU, GMT, MEX, PER, URY)
Pico	0,1–1	6	4 (ARG, BRA, PER, PRY)	5	3 (ARG, BRA, PER)
Femto	<0,1	0	N.a	0	N.a

digital television.[19] Both were financed, designed, and manufactured indigenously, the team that worked on the programmes are nationals,[20] and half the necessary pieces and all softwares are Argentinian-made. Thanks to ARSAT 1, the country became the eighth nation with the capacity to put into GEO its own satellite and the second of the American continent after the US.[21] ARSAT 2 enabled the country to retain the orbital slots allocated to it by the ITU, crucial for the provision of telecommunication services.[22]

In 2020, the Chief of the Ministers Cabinet[23] announced a new connectivity plan for Argentina, *Plan Conectar*, including amongst other initiatives a new ARSAT satellite.[24] This new system, called ARSAT-SG1 for *second generation*, will connect 200.000 rural households all over the country and beyond.[25] The service should be offered at low prices and cover rural and less populated areas,

[19] Argentina.gob.ar, "El ARSAT-1", n.d., www.argentina.gob.ar/jefatura/innovacion-publica/ssetic/conectar/el-arsat-1.
[20] Ibid.
[21] Ibid.
[22] Argentina.gob.ar, "El ARSAT-2", n.d., www.argentina.gob.ar/jefatura/innovacion-publica/ssetic/conectar/el-arsat-2.
[23] "Jefatura de Gabinete de Ministros" in Spanish.
[24] Argentina.gob.ar, "Conectar", n.d., www.argentina.gob.ar/jefatura/innovacion-publica/ssetic/conectar.
[25] Argentina.gob.ar, "El ARSAT-SG1", n.d., www.argentina.gob.ar/jefatura/innovacion-publica/ssetic/conectar/el-arsat-sg1.

Table 3 Argentinian telecommunication satellites (Gunter's Space page)

Spacecraft	Year launch	Customer/operator			Manufacturer		Orbit & mass
ARSAT 1	2014	ARSAT	Public	ARG	INVAP	ARG	GEO, 2.985 kg
ARSAT 2	2015	ARSAT	Public	ARG	INVAP	ARG	GEO, 2.975 kg
MDQubeSAT1	2022	Innova space	Private	ARG	Innova Space	ARG	LEO, <1 kg
MDQSAT-1A, MDQSAT-1B, AMAJ Mother, SatDuino	Planned 2022	Innova space	Private	ARG	Innova Space	ARG	LEO, <1 kg
ARSAT SG1	Planned 2023	ARSAT	Public	ARG	INVAP, GSATCOM	ARG, TUR	GEO, 1.900 kg

not easily connected through terrestrial infrastructure.[26] More recently, MDQube-SAT1 was launched in 2022, which is the first system of an Argentinian Long Range Wide Area Networking (LoRaWAN) pocketqube constellation, itself called *Libertadores de America.* Manufactured by the startup *Innova Space*, the network of satellites aims to democratise access to IoT connectivity everywhere, and especially in Latin America. The company plans to have coverage over the entire globe, from urban to rural areas.[27] Two to four other pocketQubes should be launched later in 2022.

4.2 Access to Telecommunication Services in Latin America

4.2.1 Adoption of Telecommunication Services

According to the World Bank, the Latin American and Caribbean region altogether had a population of 652 M in 2020.[28] As shown in Fig. 1, that same year less than 97 M people had a fixed broadband subscription (14,9%), and a little over 94 M people had a fixed telephone subscription (14,4%),[29] nevertheless 668 M subscriptions to mobile phone plans were made.[30] In the same line, Smartphones tend to be the most used device in the region; Cisco reported that, in 2017, smartphones

[26] ARSAT, "ARSAT Segunda generación", n.d., www.arsat.com.ar/satelital/satelites/arsat-segunda-generacion-1/.

[27] Alba Orbital, "PocketQube Workshop (Virtual Edition), 8 October 2020, 05:33:52, www.youtube.com/watch?v=TmpwCZwZ1qE&t=19864s.

[28] The World Bank, "Latin America", https://data.worldbank.org/region/latin-america-and-caribbean?view=chart.

[29] The World Bank, "Fixed broadband subscriptions—Latin America & Caribbean", https://data.worldbank.org/indicator/IT.NET.BBND?locations=ZJ.

[30] The World Bank, "Mobile cellular subscriptions—Latin America", https://data.worldbank.org/indicator/IT.CEL.SETS?end=2020&locations=ZJ&start=2000.

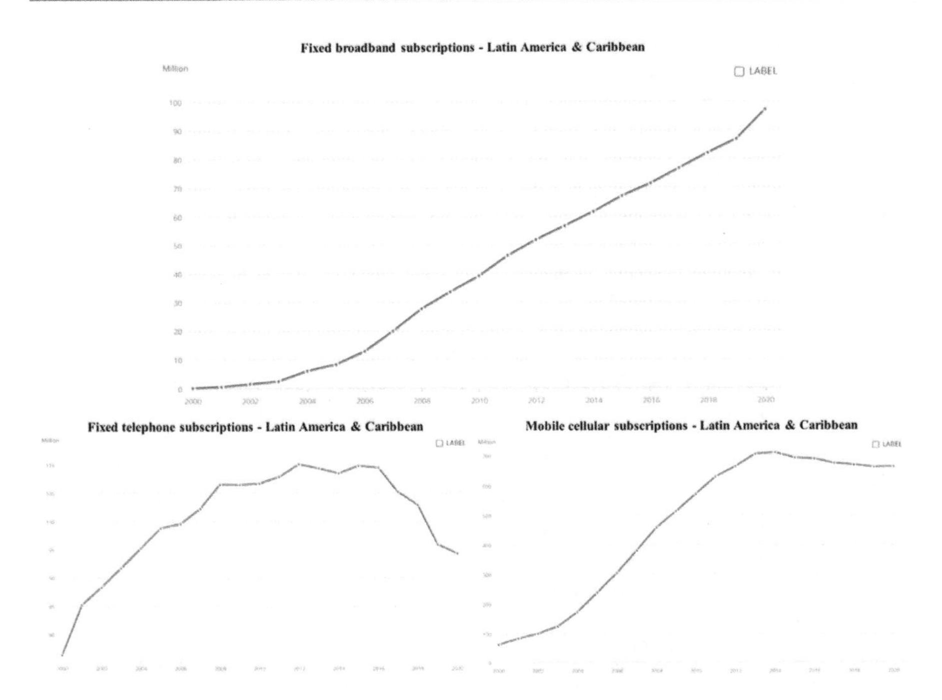

Fig. 1 Evolution of access to different telecommunication services between 2000 and 2020 in Latin America and the Caribbean (visualised by the World Bank)

were accounting for 30% of all networked devices (405.8 M), while Machine-to-machine modules accounted for 26% (352 M), non-smartphones 18% (248.3 M), PCs 12% (163.9 M), connected TVs 11% (146.9 M), and tablets 3% (36.8 M).[31]

While Fig. 1 demonstrates a general improvement over the past two decades in selected services, many rural and remote areas in Latin America are still affected by the digital gap or divide.[32] The latter phenomenon is in reference to "new forms of social inequality derived from the unequal access to the new information communications technology, by gender, territory, social class, and so forth".[33] In 2020, almost 60% of people did not have internet at home, and 26% did not have access

[31] Cisco, "VNI Complete Forecast Highlights", 2018, www.cisco.com/c/dam/m/en_us/solutions/service-provider/vni-forecast-highlights/pdf/Latin_America_Device_Growth_Traffic_Profiles.pdf.

[32] The World Bank, "Fixed broadband subscriptions", The World Bank, "Mobile cellular subscriptions", The World Bank, "Fixed telephone subscriptions—Latin America & Caribbean", https://data.worldbank.org/indicator/IT.MLT.MAIN?end=2020&locations=ZJ&start=2000.

[33] Almudena Moreno Minguez and Enrique Crespo Ballesteros, The digital divide in education in the knowledge society in Encyclopedia of Networked and Virtual Organizations (2008).

regardless of the location.[34] Of this 60% average, 49,8% were in urban areas, and a staggering 78,3% were in rural areas. Different elements might explain these numbers, such as the remoteness and/or the topography of certain locations. In this respect, in locations where technologies such as fibre optics face limitations, satellite telecommunication services can help reduce the digital divide by connecting people within the coverage of the satellite, regardless of remoteness with urban centres and topography, and as long as the necessary ground infrastructure is in place.

Nevertheless, for this measure to be effective, other factors must be considered. Increasing the coverage zones of telecommunication satellites and improving latency in the region will not be enough to secure people's access to these services if they do not have the means to do it, such as access to electricity. In this regard, the situation in rural and remote areas might be particularly challenging; in 2020, 44,8% of rural citizens in the region altogether were living in poverty and 21,3% in extreme poverty,[35] which could imply lack or no access to electricity services due to economic constraints. According to the Economic Commission for Latin America and the Caribbean (CEPAL), an average of 5,1% of people in Latin America did not have access to electricity in 2019, raising this number to 13,4% for the Indigneous communities.[36] That same year, people in rural areas were nearly 9 times more vulnerable to the lack of electricity (1,4% in urban areas against 12,2% in rural areas).[37]

4.2.2 Prices of Telecommunication Services

Latin America is an expensive place for fixed broadband. According to a 2022 study from Cable.co.uk, a subscription costs on average USD 59 per month.[38] As shown in Table 4, the Caribbean provides the most expensive subscription of the region altogether, and is the third most expensive place worldwide. For

[34] The World Bank, "Individuals using the Internet (% of population)—Latin America & Caribbean", https://data.worldbank.org/indicator/IT.NET.USER.ZS?end=2020&locations=ZJ&start=2000, CEPAL, "Statistics and Indicators—Demographic and social: Housing and basic services, Population without internet access at home, by geographic area, per capita, income quintile, and sex (percentage)", https://statistics.cepal.org/portal/cepalstat/dashboard.html?theme=1&lang=en. According to the World Data Bank, which takes into account Latin America and Caribbean countries, 74% of individuals were using the internet in 2020. This statistic refers to "Internet users are individuals who have used the Internet (from any location) in the last 3 months. The Internet can be used via a computer, mobile phone, personal digital assistant, games machine, digital TV etc." According to CEPAL, 58,4% of the population in Latin America did not have access to the internet at home in 2020.

[35] CEPAL, "Population living in extreme poverty and poverty by geographical area".

[36] CEPAL, "Population without access to electricity".

[37] Ibid.

[38] Cable.co.uk, "Global broadband pricing league table 2022", n.d., www.cable.co.uk/broadband/pricing/worldwide-comparison/#speed.

Table 4 Average price of fixed broadband per month and in USD in different regions of Latin America and the Caribbean (Cable.co.uk) for 2022[40]

Region Avg.		Most expensive countries		Least expensive countries	
Caribbean	78	British Virgin Islands	184	Cuba	22
		Turks and Caicos Islands	170	Grenada	37
		Haiti	170	Jamaica	37
Central America	44	Panama	66	Mexico	18
		Honduras	56	Ecuador	35
		Guatemala	55	Nicaragua	37
South America	55	Suriname	164	Colombia	21
		Malvinas	147	Argentina	21
		French Guiana	52	Paraguay	22

Table 5 Average cost of 1 GB of mobile data per month and in USD in different regions of Latin America and the Caribbean for 2022 (Cable.co.uk)

Region Avg.		Most expensive countries		Least expensive countries	
Caribbean	3,5	Cayman Islands	10,5	Haiti	0,4
Central America	2	Panama	3	Nicaragua	0,7
		El Salvador	1,3	Honduras	0,8
South America	4,1	Malvinas	38,5	Uruguay	0,3

comparison, in Eastern Europe the equivalent average pricing is around USD 20 (i.e. second cheapest in the world), and USD 50 in Western Europe.[39]

The situation for mobile data pricing is somewhat different, with South America taking the lead in the top three of the most expensive regions.[41] Central America does better, grabbing the ninth position. As shown in Table 5, the average cost of 1 GB of mobile data per month is half the price compared to the Southern region. The Caribbean comes sixth globally and sits in between Central and South America.

Another study from Cable.co.uk reveals an interesting fact. Telecommunication services are not only expensive in the region despite alarming poverty statistics,

[39] Ibid.

[40] Cable.co.uk considers Latin America geographically, regardless of whether some locations are independent or still semi-dependent of foreig governance. This demonstrates another level of complexity when trying to understand the impact of the space sector on Latin America; the lack of a common understanding and basis for research when we talk about the region as a whole. In this respect, places depending on foreign governance such as French Guiana will be impacted by the policies and laws of these foreign countries.

[41] Cable.co.uk, "Worldwide mobile data pricing 2022", n.d., www.cable.co.uk/mobiles/worldwide-data-pricing/.

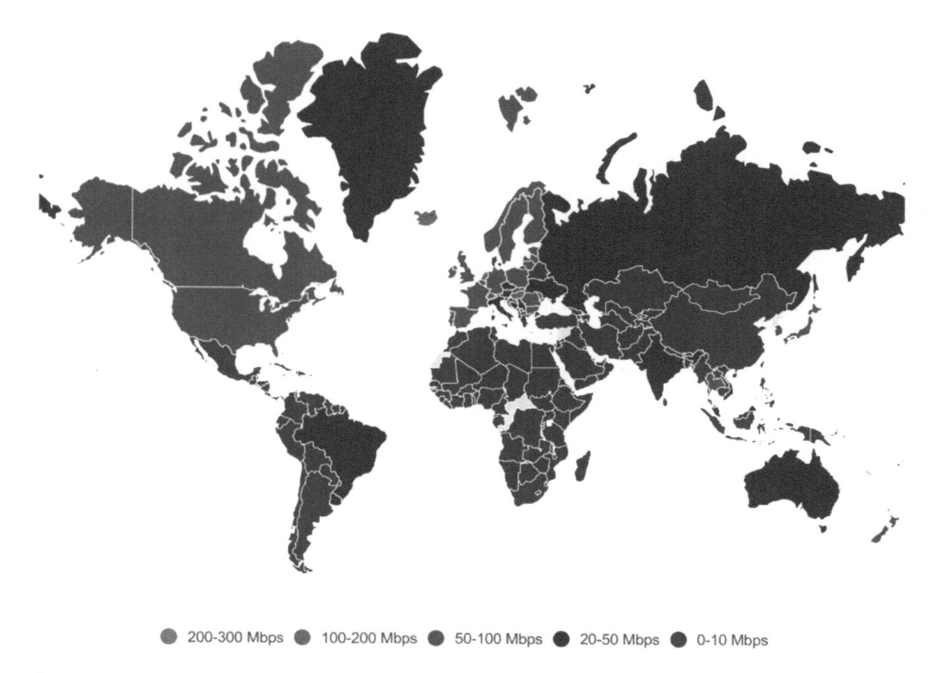

Fig. 2 Map of worldwide broadband speed in 2021 (visualised by Cable.co.uk)

but the quality of service is mediocre. The lower the Mbps, the lower the speed of the internet.[42] Broadly, and as shown in Fig. 2, Latin America seems to be running at a low speed, although some Caribbean islands present a slightly better situation compared to the rest of the region. In this respect, Aruba, Barbados, Belize, Brazil, the Bahamas, Curaçao, Grenada, Jamaica, Panama, Puerto Rico, the Cayman Islands, the Turks and Caicos islands, Trinidad & Tobago, and Uruguay are all in the fourth category where speed ranges from 20 to 50 Mbps, while all other countries in the region are in the last and slowest category.

Paying high prices for internet access is one thing, however being able to connect to the internet and charging the devices that allow for its use is another thing; again, this requires electricity, which also has a price. Out of thirteen defined regions throughout the world, the Caribbean ranks as the second most expensive place worldwide,[43] while Central America and South America ranked sixth and

[42] Cable.co.uk, "Worldwide broadband speed league 2021", n.d., www.cable.co.uk/broadband/speed/worldwide-speed-league/.

[43] According to CEPAL, this can partly be explained by the fact that access to energy is more complex in the Caribbean: CEPAL, "Sustainable Energy for all in the Caribbean, June 2016, https://repositorio.cepal.org/bitstream/handle/11362/41179/FOCUSIssue2Apr-Jun2016.pdf?sequence=1&isAllowed=y.

eighth respectively.[44] On the high end of the list, the top most expensive place is Curaçao with one kWh costing USD 0,4.[45] On the lower end, the least expensive places are Mexico and Argentina with each USD 0,05.[46]

4.2.3 Providers of Telecommunication Services

In the upstream category, one actor in the region is capable of manufacturing multi-ton telecommunication satellites: the Argentinian company INVAP. The pool of satellite operators is wider with Argentinian, Brazilian, and Mexican companies.

The downstream segment is estimated to count over a thousand companies—Table 6 lists the main one.[47] América Móvil is by far the largest actor in terms of revenues.[48] In 2020, the company generated almost USD 39B. Far behind is Telefónica with USD 18.8B. The third and fourth place go to AT&T with USD 7.4 and Millicom with USD 5.8. Other companies have generated under USD 4B. In terms of origin, América Móvil is a Mexican company, Telefónica is Spanish and AT&T American. Millicom has its headquarters in Luxembourg. Out of the four largest telecommunication service providers by revenues in the region, only one is indigneous to it. In addition, and as per 2017 numbers, Latin America has been relying mostly on US International Internet traffic, which means that "the quality of access [of Latin America] highly depends on international telecommunication infrastructure".[49]

Table 6 Main businesses present in the telecommunication sector in Latin America, per segment (Gunter's Space Page and Statista)[50]

Segment of the market	Companies
Upstream	INVAP
Midstream	ARSAT, Embratel Star One, Telecomm Mexico, Sky Mexico, DirectTV Latin America, Telebras
Downstream	América Móvil, Telefónica, AT&T, Millicom, Liberty Latin America, Telecom Argentina, Oi, TIM S.A., Entel Chile, Axtel, Algar Telecom, Empresa de Telecomunicaciones de Bogotá

[44] Cable.co.uk, "The price of electricity per KWh in 230 countries", n.d., www.cable.co.uk/energy/worldwide-pricing/.
[45] Ibid.
[46] Ibid.
[47] Crunchbase, "Latin American telecommunications companies: Summary", www.crunchbase.com/hub/latin-america-telecommunications-companies.
[48] Statista, "selected telecommunication providers".
[49] CEPAL, "State of broadband in Latin America and the Caribbean", 2017, p. 20, https://repositorio.cepal.org/bitstream/handle/11362/43670/1/S1800532_en.pdf.
[50] Statista, "Revenue generated by selected telecommunication providers in Latin America and the Caribbean in 2020", www.statista.com/statistics/998151/telecommunication-providers-revenue-latin-america/.

Lastly, according to Crunchbase, one of the lead investors in the telecommunication sector in Latin America is YC, an American startup accelerator.[51] Other lead investors are also based in the US.

5 Takeaways and Recommendations

From the information discussed throughout the document, it can be argued that space technologies can indeed support the improvement of democracy in Latin America by reducing urban–rural inequality in issues such as connectivity and telecommunication. More generally, it can bridge the digital divide faced by marginalised communities. However, and as previously mentioned, other factors must be considered for this to be effective. Regarding this, the following set of general recommendations should inform future decisions in this area.

Going small, getting closer
The major manufacturer of telecommunication satellites in Latin America is INVAP with, on average, 3-ton-satellites. Likewise, ARSAT and the other operators[52] all operate multiton systems orbiting in GEO. The region must take advantage of NewSpace and the LEO region, and this could take the form of a Starlink-style regional constellation, only smaller and more sustainable. Large GEO satellites have their advantage, but small satellites working in cohorts have many complimentary benefits to offer. They are easier and cheaper to manufacture and launch (e.g. thanks to the possibility to standardise and industrialise the manufacture, and to use Commercial off-the-shelf products), and they offer better coverage since the increased number of satellites also increases the territory covered. As they are closer to the ground, the speed should improve as well, and this would have rippling effects on other issues listed in this chapter (e.g. mediocre internet speed).

As a number of satellite constellations are being developed, the region could learn from the weaknesses (e.g. cybersecurity threats stemming from using objects that are all built the same way and that are easier to penetrate) and strengths of others to leapfrog into building its own resilient constellation.

Investing in telecommunication
As previously mentioned, lead investors tend to be from the US. In fact, in the case of YC, and perhaps others, to be integrated into the accelerator it is mandatory for the startup to be based in the US. For a region such as Latin America, where the brain-drain effect is already problematic, these sort of practices are far

[51] Crunchbase, "telecommunication companies".

[52] That includes companies who are already providing services, and excludes startups who are in the development phase of a telecom satellite or constellation such as Innova Space in Argentina.

from ideal.[53] In this sense, it is necessary to encourage local private investment through policies and programmes at the national level, promoting the establishment of local accelerators and similar hubs, and hopefully securing some sort of funding for at least the first years. Nevertheless, these initiatives may not be enough on their own. While countries in the region have different levels of economic development, and though it is not possible for each and every one of them to invest in satellite telecommunication technology individually, a grouped initiative may be the solution. Organisations such as CELAC (Community of Latin American and Caribbean States) and the newly established ALCE (Latin American and Caribbean Space Agency) could foster collective public investment. This could follow the model of the EU secured communication constellation currently under study.

Securing autonomy
Three of the four largest telecommunication companies providing services in Latin America are foreign. This makes the region dependent on others for services, as well as their prices and quality standards. Gaining the know-how to manufacture more satellites "in-house" and, perhaps, an entire regional indigebous constellation would help Latin America regain some autonomy and leeway on many variables. Multiplying upstream actors would also stimulate the development of local downstream actors. This could have repercussions on the regional ability to cover zones of its territory previously left out, and it could mean more autonomy on prices. Both of these points were mentioned earlier and identified as major issues.

Fragmenting the market
Related to the previous point, the top of the telecommunication service market seems to be concentrated on a few big actors. Likewise, foreign NewSpace actors with the potential technical capacity to provide their services onto marginalised communities (e.g. Starlink, Oneweb) are still fairly rare. Few actors on the market usually mean uncompetitive prices. By encouraging the development of satellite telecommunication, not only upstream and midstream actors can flourish and secure more autonomy for the region, but this could also allow for a number of new downstream actors using space technologies to provide services directly to end users. Sparking more competition should in turn affect the currently high prices for citizens.

Educating the users
The digital divide present in Latin America and the ensuing lower level of digital education in communities left out from the digital era could be an issue in the adoption of space services. In order to ensure that satellite telecommunication

[53] Caroline Oliveira and Hannia Guadarrama, "Brain Drain in Latin America: Why are high-skilled workers leaving? Mexico and Brazil case studies", April 2022, www.researchgate.net/publication/360117644_Brain_Drain_in_Latin_America_Why_are_high-skilled_workers_leaving_Mexico_and_Brazil_case_studies.

actually fulfils its purpose of supporting connectivity and democracy, users have to hold a certain level of digital literacy and adopt safe online behaviours. That is especially true if opening citizens to the world of information also opens the door to such issues as cyberattacks and disinformation. When drafting or updating digital transition policies, it will be up to governments to accompany developments in the telecommunication sector with education programs (that can be dispensed even in rural and remote areas).

Accessing satellite technology
As previously mentioned, increasing the coverage zones of telecommunication satellites and improving latency in the region will not be enough to secure people's access to these services if they do not have the means to do it, such as access to electricity and the necessary local ground infrastructure. In this sense, in addition to investing in telecommunication-related technologies, it is also necessary to ensure people's access to electricity and related services that allow the effective use of the benefit granted by satellite technology, especially in those rural and/or remote locations where these services are scarce or even sometimes non-existent due to, for instance, remoteness from large cities and/or widely populated sectors, and/or the presence of certain topographical features.

Thus, it is necessary that local and/or central governments not only ensure the necessary ground infrastructure for the correct use of satellite technology, but also generate policies and intervention plans to guarantee service and access to electricity in their localities, especially in rural and/or remote sectors. This may require not only the injection of greater financial resources, but also an increase in other resources to monitor compliance.

6 Conclusions and Limitations

The state of democracy in the Latin American region has shown significant tensions in recent years. Discussions on palliative measures have been common at both political and social levels, with one of these being the reduction of inequality and the democratisation of access to services to improve people's quality of life. In this context, improving connectivity and telecommunication services in rural and/or remote areas has also been widely highlighted as a necessary measure to tackle territorial inequality and to fulfil the democratisation of access to services, with the space sector and satellite development being an opportunity for this.

As argued throughout the document, space technologies can indeed support the improvement of democracy in Latin America by reducing urban–rural inequality in issues such as connectivity and telecommunication, however, other factors must be considered for this to be effective. As reviewed throughout this document, it is not enough to increase the coverage zones of telecommunication satellites and improve latency in the region, but it is also fundamental to secure people's access to electricity and the necessary local ground infrastructure. Thus, before investing in and ensuring the availability of satellite technology for this purpose, it would be necessary to generate both policies and monitoring measures to ensure full access to the benefits of such technology.

Considering the above, to facilitate the use of satellite technology and its benefits for the aforementioned purposes in the region, it is necessary to consider aspects related to economics and regional cooperation. In particular, although it would indeed be ideal to encourage local private investment, the different levels of national economic development in the region do not necessarily allow this to be a viable option, so grouped initiatives from organisations such as CELAC and ALCE could make the development and use of this technology a more economically viable possibility, facilitating access to it and its benefits in the different latinamerican countries.

Finally, despite the information and recommendations described in this article, further research on this topic is required. Specifically, this document should be used as a preliminary collection of information on the benefits of space sector development in such diverse and important areas as the improvement of democracy-related aspects and people's quality of life. In this respect, as this review focuses on the Latin American context as a whole and does not cover the situation of each country individually, it would be necessary to explore in future reviews how the space sector and satellite technology could contribute to the improvement of these topics at the specific national level, describing in detail the situation of each country in this regard. In addition, it should also be mentioned that, due to the lack of information and the non-existence of a centralised source of data, in this paper we use information obtained from various sources, of which some incorporate the Caribbean as a Latin American sub-region, others only incorporate the subregions of Central and South America, and others only consider the countries of the region where data is available. In this sense, there may be some inconsistencies in the review presented here, so a more localised review would also be necessary in future articles.

Lauryn Lee Hallet holds an LL.M. in International Law from the University of Bristol. She is a space law and policy researcher, and recently became a cybersecurity consultant. She has published several research on space-related topics, including on the legal regime around satellite constellations, the impact of constellations on astronomy, and on the benefits of space for the blue economy. Other areas of interest include cybersecurity and disinformation.

Marieta Valdivia Lefort is a Ph.D. candidate at the University College London (UCL) Institute of the Americas. In addition, she holds an MSc in Public Policy from the UCL Department of Political Science, School of Public Policy. Her research interests include policy studies, politics, ideologies and institutions, regulation and public value, social and political behaviour, and diplomacy.

Social Sustainability: A Challenge for the Supply Chain of the Mexican Space Sector

Lisette Farah-Simón, Miguel Angel Reyna-Castillo, and Hugo Javier Buenrostro-Aguilar

Abstract

Space sustainability is a challenge that goes beyond conflicts in space; it must also be taken into account within the processes that are done on Earth, such as the manufacture of mechanisms. The literature indicates that it is not enough to take care of carbon footprint; the social impact of these processes must also be considered, with special attention to the so-called emerging economies. Therefore, this chapter aims to analyse possible paths for social sustainability practices in the supply chain of the Mexican space sector. This exploratory approach was carried out on the basis of a conceptual analytical methodology, intended to specify variables and put forward a theoretical model. This paper contributes to fill the gap in the literature on social issues in downstream suppliers in the emerging Mexican space sector. A measurement model based on work in emerging economies is proposed for the evaluation of social sustainability in the supply chain of the Mexican space sector.

L. Farah-Simón (✉) · H. J. Buenrostro-Aguilar
Universidad Nacional Autónoma de México, Ciudad de México, México
e-mail: farah@comunidad.unam.mx; yroxaber@hotmail.com

M. A. Reyna-Castillo
Universidad Autónoma de Tamaulipas, Tampico, Tamaulipas, México
e-mail: mreyna@docentes.uat.edu.mx

A. Froehlich (ed.), *Space Fostering Latin American Societies*, Southern Space Studies,
https://doi.org/10.1007/978-3-031-20675-7_4

1 Introduction

The world's productive sectors have received a constant call to join and contribute to sustainable development where economic growth coexists with care for the environmental and social dimensions. So far, the social dimension has received less attention in the literature and in practice, particularly in developing countries.[1,2] Given that the world's raw materials and supply are mainly developed and produced in emerging economies, it has been considered strategic and urgent to boost the social care of these sectors by having the focal companies put pressure on their supply chains to require evidence of social sustainability policies. Initiatives in developed economies such as Switzerland and Germany have seen the role of the focal purchasing company as an intermediary as a key factor in promoting social sustainability in emerging economies.[3] However, as Yawar and Seuring point out,[4] corruption, especially in developing countries, is evident and suppliers can easily avoid sanctions by paying bribes to the government. Reflection on social issues in the chain has become a pressing issue, as recurring problems have been seen in relation to working conditions, child labour, human rights, health and safety, development of minorities, and inclusion of disabled/marginalised people and genders.[5,6]

Social problems in emerging markets are, therefore, certainly a matter of ethical concern, especially in the context of the Covid-19 pandemic, where conditions have made workers vulnerable in terms of labour and human rights, due to the constant risk of employers seeking the survival of their businesses at the expense of the rights of their employees.[7] In addition to being a matter of ethical implication, social sustainability is urgent in the productive sense, as its relationship with the operational performance of the supply chain has been demonstrated, including with the financial performance of companies.[8]

---

[1] Jayani Ishara, Sudusinghe and Stefan Seuring, "Social sustainability empowering the economic sustainability in the global apparel supply chain", *Sustainability*, 12 (2020).

[2] Mohammad El Wali, Saeed Rahimpour Golroudbary and Andrzej Kraslawski, "Circular economy for phosphorus supply chain and its impact on social sustainable development goals", *Science of The Total Environment*, 777 (2021).

[3] *Ibid.*

[4] Sadaat Ali Yawar and Stefan Seuring, "Management of social issues in supply chains: a literature review exploring social issues, actions and performance outcomes", *Journal of Business Ethics* 141, No. 3, (2017): 621–643.

[5] Ibid.

[6] Deniz Köksal, Strähle, Müller and Freise, "Social sustainable supply chain management in the textile and apparel industry - A literature review", *Sustainability* 9, No. 1 (2017): 100.

[7] Alexander Trautrims, Martin C. Schleper, M. Selim Cakir and Stefan Gold, "Survival at the expense of the weakest? Managing modern slavery risks in supply chains during COVID-19", *Journal of Risk Research*, 23 No. 7–8, (2020): 1067–1072.

[8] Vishnu Nath and Rajat Agrawal, "Agility and lean practices as antecedents of supply chain social sustainability", *International Journal of Operations & Production Management* 40, No. 10. (2020): 1589–1611.

Although the audit and practice of social sustainability is a pending agenda for the productive sectors, this is linked to the challenge that academia also faces in terms of defining models and indicators that really explain social welfare in the links of the supply chain. The question of what a supplier company should consider to ensure that it is taking care of the people associated with it (employees, customers, stakeholders, macro-social[9]) has not been fully answered. The space sector has not been oblivious to this call and this search. One of the trends in the global space sector is precisely the integration of sustainable practices in its activities. The aim is that, from a sustainable development perspective, space innovation and technological development contribute to the reduction of negative externalities in the future of space activities. Incorporating sustainability in operations will allow us to continue using the space environment for the development of activities, such as: national security, Earth observation, telecommunications, data transfer, navigation, scientific exploration, economic development, among others.[10]

It is a reality that the concept of space sustainability has become more relevant in the space community, to the point that the United Nations Office for Outer Space Affairs (UNOOSA) has formed a working group on Long-term Sustainability of Outer Space Activities, which addresses thematic areas such as: space debris; space operations and tools to support collaborative space situational awareness; space weather; regulatory regimes and guidance for actors in the space arena.[11] The Report of the Scientific and Technical Subcommittee on its 56th session, held in Vienna from 11 to 22 February 2018, states the conclusion of the working group in charge of the draft structure of the Space2030 Agenda, which mention the importance of

> promoting, through international cooperation, the use of space technologies and data for sustainable economic and social development in developing countries by strengthening the capacity to use space technology or strengthening capacity; increasing their leaders' awareness of the cost-effectiveness and complementary benefits that can be derived from such technologies and data; and increasing outreach activities to raise awareness of these benefits.[12]

[9] Sonu Rajak and Sekar Vinodh, "Application of fuzzy logic for social sustainability performance evaluation: a case study of an Indian automotive component manufacturing organization", *Journal of Cleaner Production*, 108, (2015): 1184–1192.

[10] Committee on the Peaceful Uses of Outer Space [COPUOS], *"Proposal for a Draft Report and a Preliminary Set of Draft Guidelines of the Working Group on the Long-term Sustainability of Outer Space Activities: Working Paper by the Chair of the Working Group"*, Vienna (2013), 2, http://www.unoosa.org/pdf/limited/c1/AC105_C1_L339E.pdf (all websites cited in this publication were last accessed and verified on 07 July 2022).

[11] United Nations Office for Outer Space Affairs [UNOOSA], *"Long-term Sustainability of Outer Space Activities"*, https://www.unoosa.org/oosa/en/ourwork/topics/long-term-sustainability-of-outer-space-activities.html.

[12] *Ibid.*, 14.

In addition, this subcommittee mentions the importance of joint efforts of global partnerships and increased cooperation of Member States; United Nations entities; intergovernmental and non-governmental organisations; industry; and private sector entities.[13] It is important to highlight that the efforts of the space industry are grouped in three sectors[14]:

- Civil: Non-defence-government space activities, including launching and managing satellites, conducting research and exploring the solar system.
- National Security: This sector includes the defence and intelligence sectors. It refers to activities that are involved in the operation of space assets for intelligence purposes to support military and law enforcement operations.
- Commercial: All space-related endeavours—including goods, services and activities—provided by private sector enterprises with the legal capacity to offer their products to non-governmental customers. Commercial space efforts range from satellite communication to space tourism.

At present, the study of space sustainability has focused on activities that take place outside the Earth's atmosphere; however, there is a large number of actions that occur on the ground and that support these activities, from design to production, launching and putting satellites into orbit involving a variety of actors that also determine the sustainability of the sector. In view of this, it is essential to study the aerospace supply chain, which is characterised by several gaps in its performance, because, although they have been significant, they still only focus their efforts on environmental issues. For example, the recent study by Paladini, Saha and Pierron,[15] who, based on the ESA-MELiSSA model on sustainability and circular economy, propose its terrestrial adaptation for water treatment; or the proposal of recyclable additive material for 3D-construction in space by Ishfaq et al.[16] However, it is necessary to retake aspects that add up to the integral vision that sustainability requires, such as the social aspect, where there is still a lack of indicators that can be applied in this manufacturing sector.

An emerging market that has played a key role as a link in global space sector chains is Mexico. According to information generated by the Ministry of Economy,[17] the Mexican aerospace sector has positioned itself nationally as the fourth

[13] *Ibid.*, 64.

[14] Space Foundation, "*Space Briefing Book: Space Sector*", https://www.spacefoundation.org/space_brief/space-sectors/.

[15] Stefania Paladini, Krish Saha and Xavier Pierron, "Sustainable space for a sustainable Earth? Circular economy insights from the space sector", *Journal of Environmental Management* 289, No. 1, (2021): 112, 511.

[16] Kashif Ishfaq, Muhammad Asad, Muhammad Arif Mahmood, Mirza Abdullah and Catalin Iulian Pruncu, "Opportunities and challenges in additive manufacturing used in space sector: a comprehensive review", Rapid Prototyping Journal, (2021).

[17] Secretaría de Economía, "Plan Nacional Estratégico de la Industria Aeroespacial", p. 38.

most relevant after the food industry, the automotive industry and the electric-electronic sector; likewise, it has become an active supplier of levels one and two. Based on the above, this paper aims to analyse possible paths for social sustainability practices in the supply chain of the Mexican space sector. This exploratory approach was based on a conceptual analytical methodology, intended to identify variables and possible relationships within a model. This contributes to fill the gap in the literature on social issues in downstream suppliers in the emerging Mexican space sector. An overview of the industry's supplier trade is presented, as well as a vision of the advances in sustainability in the aerospace sector. A model of measures based on studies of emerging economies is proposed to integrate and establish a structure that encompasses measures related to the social aspects of sustainability in the emerging Mexican space sector. Lastly, conclusions are drawn.

2 The Aerospace Supply Chain in Mexico

Addressing the analysis of the supply chain of any economic sector implies recognising the activities involved from the procurement of raw materials, their transformation, to their transfer to the end consumer. This has to be subject to the new needs of customers, without forgetting the socio-environmental setting. So, it is essential that the supply chain is committed and performs actions that favour the achievement of sustainability, not only in outer space but also in the large number of activities that are performed on the ground to obtain the final products of this sector, which is of such relevance today. Mexico, as one of the main international suppliers, must join these efforts, otherwise it will lose competitiveness, which in turn creates significant challenges that require integral and transversal solutions from a sustainable perspective. In this sense, it is essential to recognise that, in general terms, the aerospace supply chain provides the raw materials and components needed for the manufacture, maintenance, repair and overhaul of a wide range of elements necessary for the space sector.[18] This chain is led by producers, structured at different levels[19]:

- Original Equipment Manufacturers (OEMs), which are in charge of end products for the consumer market and which, in relation to the aerospace sector, focus on design, development and manufacturing.
- Tier 1: direct suppliers of parts to OEMs, thus becoming sub-manufacturers or assemblers of sections and systems including engines, avionics systems, aircraft interiors, landing gears, etc.

[18] ITF, "La cadena de suministro aeronáutica", *Hojas informativas producidas por la Federación Internacional de los Trabajadores del Transporte*, https://www.itfglobal.org/sites/default/files/resources-files/scalop_aeronautical_supply_chains_spanish.pdf, 1.
[19] Ivan Koblen and Lucia Nižníková, "Selected aspects of the supply chain management in the aerospace industry", *Incas Bulletin* 5, no. 1 (2013), 137.

- Tier 2: key Tier 1 suppliers, focusing on aerostructures, subsystems and subassemblies.
- Tier 3: broader suppliers specialising in the production of components and specific processes related to raw materials and electronic components.

This generates a series of interrelationships between each level, or link, which must be analysed and understood in order to observe their value and importance within the sector, as well as their economic, environmental and social effects on the chain itself. In this sense, it is possible to study the manufacturing processes of the space industry within four major areas—(1) product modification, (2) product improvement, (3) process modification and (4) process improvement—where products are the tangible part (what is done) and processes are the intangible part (how it is done).[20] Both elements (tangible and intangible) have to face a variety of challenges and pressures to reduce costs, increase productivity and competitiveness under a responsible approach that rethinks the structure and purpose of the chain itself; this creates the right environment for innovation and the adoption of best practices that lead the sector, its manufacturers, suppliers and customers to implement correct business strategies focused on achieving objectives.[21]

This has not been far away in the development of the Mexican aerospace supply chain, which has an estimated market of US$ 450 billion, with a high share in the global supply chain, made up of 238 companies located in 17 states, which create 29.000 jobs and have exports worth over US$ 3 billion and imports worth over US$ 2 billion[22]; this has allowed this sector to become one of the most relevant for the development of the country due to its economic and social potential.

Thus, the Mexican aerospace industry has grown steadily since 2004, which has positioned Mexico as the 14th largest international supplier, due to the fact that its activities include design, manufacturing and marketing with a complex supply chain that includes maintenance, repair and rehabilitation services; suffice it to mention that in 2019 only the manufacture of aerospace equipment accounted for 0,13% of GDP, which translates to US$ 3,446 billion.[23] These results have been achieved thanks to a network of suppliers, manufacturing companies and

[20] Yuri Romaniw and Bert Bras, "Survey of Common Practices in Sustainable Aerospace Manufacturing for the Purpose of Driving Future Research", presentation at the 19th CIRP International Conference on Life Cycle Engineering, Berkeley, USA, p. 3, 23–25 May 2012.

[21] Koblen and Nižníková, "Selected aspects of the supply chain management in the aerospace industry", pp. 135–136.

[22] Secretaría de Economía, "Plan Nacional Estratégico de la Industria Aeroespacial", p. 10, https://www.gob.mx/cms/uploads/attachment/file/58802/Plan_Estrat_gico_de_la_Industria_Aeroespacial_junio.pdf.

[23] Jorge Olmilledo and Lucía Rabanal, "Sector aeroespacial en México", *ICEX*, pp. 1–2, 31 December 2020, https://www.icex.es/icex/GetDocumento?dDocName=DOC2021873961&urlNoAcceso=/icex/es/registro/iniciar-sesion/index.html?urlDestino=https://www.icex.es:443/icex/es/navegacion-principal/todos-nuestros-servicios/informacion-de-mercados/estudios-de-mercados-y-otros-documentos-de-comercio-exterior/ficha-sector-aeronautico-mexico-2020-doc2021873961.html&site=icexES.

infrastructure that have formed clusters. From 1999 to Q3 of 2020 mainly five Mexican states contributed to it[24]:

- Queretaro, which amassed an investment of US$ 989,71 million, with the participation of several companies such as Bombardier, Aernova, Airbus, ETU División Aeronáutica, Global Composites Manufacturing, HYRSA, ITP Aero, RYMSA or PCNC, which focus on assembly, manufacture of aircraft or aeroparts, turbojets, landing gear and parts, and goods for the repair or maintenance of aircraft and parts.
- Baja California, which amassed an investment of US$ 716,69 million, from companies such as Honeywell, Safran Group, Collins Aerospace, Gulfstream and Lockheed Martin, which focus on innovation in aircraft interior design, as well as complete integration tests, precision machinery, electrical and hydraulic systems and metal plate forming processes.
- Chihuahua received investments of US$ 682,56 million, with companies such as Beechcraft, Bell Helicopter, Honeywell, Ez Air, Kaman Aerosystems, Safran Group, Soisa Aircraft Interiors and Textron Aviation, focused on high-precision machinery, aeroparts, harnesses, equipment for emergency landings at sea, seats, and heat and chemical treatments for metals.
- Nuevo León received US$ 341,26 million, with companies such as Aero Alterno, Azor, Aerovitro, Aeisa, Cimsamex, Hemaq, Mimsa and Quimmco.
- Sonora amassed an investment of US$ 238,9 million mentioned, with companies such as Goodrich, Rolls-Royce, Esco, Radiall, Williams International, Incertec, Bodycote, Ducommun AeroStructures, Groupe Latécoère or Daher, focused on the manufacture of turbines, casting processes, machining and secondary processes, aerostructures and composite materials, among other activities.

Thus, it can be said that the Mexican North West region specialises in electrical-electronic elements, while the North Central region focuses on component assemblies and repair and maintenance activities. 43% of the companies are medium-sized and employ between 51 and 250 workers; of the total number of companies, 80% are manufacturing companies, and the rest are divided into various activities such as design and engineering, maintenance, repair and supervision.[25] Through the institutionalisation of the National Strategic Plan for the Aerospace Industry, the Mexican federal government seeks to improve current conditions through the implementation of five strategies[26]:

[24] Jorge Olmilledo and Lucía Rabanal, "Sector aeroespacial en México", pp. 4–5.
[25] Secretaría de Economía, "Plan Nacional Estratégico de la Industria Aeroespacial", p. 41.
[26] Ibid., pp. 3–5.

- Promoting and developing the domestic and foreign markets, with the aim of defining new niches in the supply chain and innovation networks; taking advantage of domestic purchases for the development of the sector; and participating in international projects and programs.
- Strengthening and developing the capabilities of the domestic industry, with the objective of having a developed and integrated supply chain; achieving a comprehensive approach to the complete life cycle of the product, incorporating design, engineering, manufacturing, conditioning and repair; developing current and new clusters by identifying regional vocations, attracting strategic suppliers, generating collaboration between OEMs and Tier 1, in a scheme in which the private sector, the federal government and state governments collaborate, with robust logistics and financing schemes; developing capacities for assembly with high technological content; facilitating the internationalisation of companies established in the country; developing technology; achieving specialisation in processes or products; and developing infrastructure.
- Developing human capital: of great relevance is the education, training, specialisation and technical assistance for specialists within the aerospace sector, which is sought to be achieved with joint programs; developing technical and specialised careers; human resources training; and a link between academia and industry.
- Necessary technological development, with the promotion of Technological Development Centres participating in the sector.
- Developing cross-cutting factors, i.e., an institutional framework; government leadership; support programs; enabling and incentivising regulation; infrastructure; and international agreements.

These five strategies are far-reaching and reveal the interest in projecting the country's aerospace sector by creating actions that, transversally, promote the development of this sector, all of which requires a true political will to create good relations with academia and the private sector in order to better position the national sector, for which it is essential to overcome challenges and improve current conditions in order to achieve better competitiveness committed to the sustainability of the sector.

3 Sustainability and Challenges of the Mexican Aerospace Chain

The global value chain of the space industry is complex, due to its multiple dynamics that require a holistic view considering multiple specifications, such as the following: a constantly evolving regulatory and policy environment, multilateral collaboration, dependence on cross-cutting macro-trends, implications with other industrial sectors, and considerations of social, economic and environmental impact.[27] As space is a global and limited resource, its use and exploitation is based on the principles of freedom and equality, peaceful utilisation, prohibition of claims to its sovereignty, and international cooperation.

Space sustainability is related to the threat posed by space debris, space weather, radio frequency interference and objects in Earth orbit; and with that, some initiatives have been put forward to solve these problems. The Secure World Foundation (SWF) has defined space sustainability as "*Ensuring that all humanity can continue to use outer space for peaceful purposes and socioeconomic benefit now and in the long term*".[28]

Efforts to promote space sustainability have evolved since the stakeholders involved have understood that any action they take can have a long-term impact on future space activities. However, the concept of space sustainability is still very ambiguous, since the idea of implementing a model that guarantees the use of outer space has become widespread. Without establishing the sustainable needs of the processes carried out on Earth to achieve the successful development of a space mission, and thus contribute to the fulfilment of the SDGs in a comprehensive manner.

In Mexico, with the approval of the law creating the Mexican Space Agency (AEM) in 2010, it was established that the Agency would be in charge of designing and executing space policy in Mexico, as well as promoting manned space transportation programs, the satellite area, launch services, developing new technologies and promoting research and education in space science and technology. The approval of this law requires the strengthening of the space industry in Mexico, as well as the development of a national strategic plan that includes, in all its guidelines, aspects of sustainability, social responsibility and environmental protection.

The positioning achieved by the national aerospace sector still suffers from situations that must be overcome in order to achieve economic objectives with lower environmental and social impact. In this sense, there are evidently major challenges that can be categorised as internal and external factors of the national supply chain.[29] Within the efforts to achieve sustainability in the supply chain, in this case aerospace, through a triple bottom line, the need to take actions in relation to the following areas is evident[30]:

[27] PwC Space, "*Main trends and challenges in the space sector*", France, (2019). 4–5.

[28] Secure World Foundation, "*Space Sustainability: A practical guide*", California (2014), 4 https://swfound.org/media/121399/swf_space_sustainability-a_practical_guide_2014__1_.pdf.

[29] Jennifer Orozco, Beatriz Rosas and Zulema Córdova, "Desarrollo de proveedores en la cadena de valor: La Industria Aeroespacial en Mexicali, Baja California, México", *Sinapsis, Revista de Investigaciones de la Institución Universitaria EAM* 11, no. 2 (2019), 36–39.

[30] Kavitha Gopalakrishnan, Yahaya Yusuf, Ahmed Musa, Tijjani Abubakar and Hafsat Ambursa, "Sustainable supply chain management: A case study of British Aerospace (BAe) Systems", *Int. J. Production Economics* 140, (2012), 200–201.

- Integration of the supply chain, which will allow to build parameters towards sustainability; provide value to suppliers; choose suppliers that comply with environmental and social standards; carry out audits with criteria based on quality, flexibility, environmental policies and compliance with the legal framework.
- Methods to reduce costs, through financial audits that review economic progress; adoption of techniques that increase operational efficiency; and a constant review of the progress of suppliers, employees and facilities in meeting sustainable goals.
- Quality and safety protocols to control the quality of operations, personnel and product safety; adoption of safety protocols for the welfare of employees; in the case of subcontracting companies, diagnosis of the compliance of quality and safety protocols by suppliers is essential.
- Reduction of carbon emissions throughout the supply chain; an increase in emissions would lead to chain inefficiency, so carbon footprint must be calculated to achieve a balance between emissions and costs through maximisation of logistics services.
- Check that the supply of raw materials is sustainable, exploring the availability of resources and distances; it is also a priority to design, evaluate and operate mechanisms that allow the substitution of non-renewable resources for renewable and recyclable options, giving preference to materials manufactured in an efficient and sustainable manner.
- Government legislation, which must be known in order to comply with environmental and social standards and regulations.
- Existence of a department focused on ensuring the achievement of social, environmental and economic goals, as an exclusive unit made up of experts and managers to guarantee chain sustainability.
- Organisational culture and employee participation, allowing the creation of a set of organisational values that include sustainability; for this purpose, it will be necessary to organise awareness and training campaigns on related topics, promoting an outward-looking culture that serves the organisation's internal and external community.
- Identify key performance indicators that evaluate essential areas in terms of factors and challenges that affect the achievement of sustainable goals, with comparative evaluation of progress that ensures not only environmental and traditional economic benefits, but also seeks social returns.

Undoubtedly, the social aspect is the least studied and least in depth within the action plans, so a major effort is required to integrate social issues when designing, implementing and evaluating the supply chain. This should occur in two stages, the first within the organisation, including employees and suppliers; and the second outside, observing the impact on the community, which rather than being a foreign agent is a key player to assess the perception and symbiosis of the organisation with its environment.

4 Pathways for Social Sustainability in the Space Industry: Approaching a Model

The academic debate on the fusion of social sustainability and supply chain management has increased in the last decade and its integrated measurement of global supplier performance has been recognised as an emerging path in the discourse of Sustainable Supply Chain Management.[31] Regional and global social sustainability in this area is a pending issue as emphasised in the Sustainable Development Goals of the United Nations 2030 Agenda[32] especially from the global sourcing in so-called emerging markets and developing economies.[33]

Among the emerging countries that have stood out for their concern to define, evaluate and expose indicators related to sustainable social management of the supply chain in manufacturing sectors is India. The work of Rajak and Vinodh[34] modelled a set of indicators to assess social sustainability performance in Indian manufacturing companies. They propose four corporate enablers to classify social sustainability indicators: the company's internal society (internal human resources), external population (immediate community to the company), stakeholders (partners, corporate governance, suppliers) and macro-social performance (actions advanced by the company). An outstanding value of this research is the structure for ordering corporate social aspects. Another of the initial works, in emerging economy, was carried out in Bangladesh, where social sustainability practices in manufacturing companies of specific lines of business were explored, as was the case of the study of three companies involved in oil, gas and the other dedicated to the manufacture of tires. In this work, Mani et al.,[35] empirically validated their measures and identified several social sustainability practices, resulting in a reliable valid scale of 20 items under several social dimensions: health and safety, bonded child labour, employee education, wages, human rights, gender diversity and discrimination, sanitation, philanthropy, engagement, community employment, as social issues relevant to the manufacturing sector.

[31] Jayani Ishara Sudusinghe and Stefan Seuring, "Social sustainability empowering the economic sustainability in the global apparel supply chain", *Sustainability* 12.7 (2020): 2595.

[32] El Wali, "Circular economy for phosphorus supply chain and its impact on social sustainable development goals", p. 3.

[33] Nikolas K Kelling, Philipp C. Sauer, Stefan Gold and Stefan Seuring, "The role of institutional uncertainty for social sustainability of companies and supply chains", *Journal of Business Ethics* 173, No. 4, (2021): 813–833.

[34] Rajak and Vinodh, "Application of fuzzy logic for social sustainability performance evaluation: a case study of an Indian automotive component manufacturing organization", 1187.

[35] Venkatesh Mani, Angappa Gunasekaran and Catarina Delgado, "Enhancing supply chain performance through supplier social sustainability: An emerging economy perspective", International *Journal of Production Economics* 195 (2018): 259–272.

Also, from the Asian approach, the study by Ahmadi et al.[36] used the best-worst-method to validate social sustainability measures with 38 supply chain experts from different lines of business in Iranian companies. Eight measures of social sustainability relevant to the industrial sector were obtained: occupational health and safety, training education and community influence, contractual stakeholder influence, occupational health and safety management system, employee interests and rights, stakeholder rights, information disclosure and employment practices.

The work of Nath and Agrawal,[37] following up on the proposal of Donna Marshall et al.,[38] studied Indian manufacturing suppliers and validated indicators considering three levels of adoption of social sustainability in the supply chain: On the first level they placed the orientation towards social sustainability of the focal company, where aspects related to awareness, importance and agreement with the importance of social sustainability were considered. On a second level, they proposed the Basic Practices, where they considered as indicators (1) Periodic supervision of suppliers in terms of health and safety standards, (2) Design of systems for work-life balance throughout the chain, (3) Design of a code of ethics for suppliers, and (4) Verification that there is inclusion and gender equity in suppliers. And on the third level, Advanced Practices, where the following items were considered: (1) Development and innovation of products that reduce risks to the health of consumers, (2) Development of processes that have reduced the danger to the health and safety of workers, and (3) Development of processes with partners to benefit employees, community groups and fair trade with suppliers throughout the chain in the long term. Recently, Kottala's research[39] validated influential factors in the enactment of social sustainability in the supply chain activities of the Indian manufacturing sector as viewed from a social development perspective. The factors resulting from their research were linked to community, safety, product responsibility and sustainable business opportunities.

Also, recently, Sudusinghe and Seuring[40] proposed a measurement model where they categorised the social aspects of social sustainability in the supply chain in the garment manufacturing sector in Sri Lanka. Among the aspects they

[36] Hadi Badri Ahmadi, Simonov Kusi-Sarpong and Jafar Rezaei, "Assessing the social sustainability of supply chains using Best Worst Method", *Resources, Conservation and Recycling* 126 (2017): 99–106.

[37] Vishnu Nath and Rajat Agrawal, "Agility and lean practices as antecedents of supply chain social sustainability", International Journal of Operations & Production Management, pp. 1610–1611.

[38] Donna Marshall, Lucy McCarthy, Paul McGrath and Marius Claudy, "Going above and beyond: how sustainability culture and entrepreneurial orientation drive social sustainability supply chain practice adoption", *Supply chain management: an international journal,* 20 No. 4 (2015): 434–454.

[39] Sri Yogi Kottala, "Social Sustainable Supply Chain Practices Evidence From the Indian Manufacturing Sector: An Empirical Study". *International Journal of Social Ecology and Sustainable Development (IJSESD)* 12, No. 2, (2021): 73–98.

[40] Jayani Ishara, Sudusinghe and Stefan Seuring, "Social sustainability empowering the economic sustainability in the global apparel supply chain", p. 18.

propose, they emphasise care for external society: youth employment, philanthropy, disaster and emergency response. As well as the consideration of social issues in the internal care of the supplier company: education and training of employees, inclusion, gender equity, fair salary, health and safety, ethics, abolition of child labour, anti-corruption policies, among others. Their contribution offers a restructuring of the items, as well as the incorporation of some other items not included.

Latin America has also approached sustainability research in a discreet manner, for example, that carried out in industrial cooperatives in Ecuador, Peru, Guatemala and El Salvador. The study by Rodríguez, Giménez and Arenas[41] linked social sustainability with value creation. The proposed theoretical model was validated with a case study involving a project led by an NGO. This project carried out supplier development programs for poor suppliers in cooperation with several companies. The measures of social sustainability used were three dimensions of social capital: trust and mutual understanding between the company and the NGO, communication channels between companies and NGOs, and connections developed throughout the initiative.

5 Results: Elements of the Model

Returning to important works in the literature, this model has three fundamental vertices:

- Contextualisation integrity: takes into account fundamental contexts that relate to the supply chain.
- Integrity of the moments of sustainability integration in a supply chain: part of the business pedagogical reality where firms sustain a learning curve that goes from awareness to practice and from practice to innovation. In other words, it moves from knowing to doing and from doing to proposing.
- Integrity of the essential aspects of human welfare: There is no socially sustainable business development without gender equity, fair wages, health, labour welfare, governance and collaboration.

[41] Jorge A Rodríguez, Cristina Giménez and Daniel Arenas, "Cooperative initiatives with NGOs in socially sustainable supply chains: how is inter-organizational fit achieved?", *Journal of Cleaner Production* 137 (2016): 516–526.

5.1 Contextualisation Integrity

Taking up Rajak and Vinodh's structure for the macro dimensions of social sustainability,[42] the first element of the model was established as the integration of three of the four contexts proposed in their work:

- The internal company (internal human resources of the supplier company).
- External population (the company's immediate community where the customer can be included).
- Stakeholders (the involvement and care of the fair relationship between partners, members of the corporate governance and suppliers).

5.2 Integrity at Points of Incorporation in a Supply Chain

It is considered that the distinction systematised by Donna Marshall et al.,[43] and continued by Vishnu Nath and Rajat Agrawal,[44] is relevant, where the macro aspects already pointed out in Rajak and Vinodh are systematised,[45] pointing out how social sustainability is not a black-and-white dualism in terms of its assimilation within supply chain management, rather there is a progressive incorporation, where initially management and middle management emerge from lack of knowledge on the subject, indifference and even sceptical antagonism. The authors propose a progressive evaluation of social sustainability in the supply chain, at three levels of incorporation:

- ***First level***: Orientation towards social sustainability. Awareness and conviction about the importance for the social and business good (Table 1).
- ***Second level***: Basic social sustainability practices. From culture, we move on to practice and at this level actions are already taken to take care of social aspects in supplier companies (Table 2).
- ***Third level***: Advanced social sustainability practices. The supplier company no longer only incorporates and takes care of aspects of its internal and external society and stakeholders. Now the company itself is restructured and goes

[42] Rajak and Vinodh. "Application of fuzzy logic for social sustainability performance evaluation: a case study of an Indian automotive component manufacturing organization", 1187.

[43] Marshall, McCarthy, McGrath and Claudy. "Going above and beyond: how sustainability culture and entrepreneurial orientation drive social sustainability supply chain practice adoption", 440–441.

[44] Nath and Agrawal. "Agility and lean practices as antecedents of supply chain social sustainability". International Journal of Operations & Production Management, 1592–1593.

[45] Rajak and Vinodh. "Application of fuzzy logic for social sustainability performance evaluation: a case study of an Indian automotive component manufacturing organization", 1187.

Table 1 Proposed indicators for the Level 1 construct: Orientation towards SS

Dimension and item by level of incorporation of social sustainability (SS)	Internal society (human resources)	Immediate external society (community)	Stakeholders (partners, corporate order, suppliers)
Level 1. Orientation towards SS			
– Information is provided to all employees to understand the importance of social sustainability	X		
– Awareness in senior management of socially sustainable development	X		
– Social sustainability is a core corporate value in their company			X
– Seeking to promote social sustainability as a major objective in all departments	X		
– There is a clear policy statement calling for social sustainability in all areas of operations	X		
– Social sustainability is a high priority activity in their company	X		
– The company has a responsibility to be socially sustainable	X		
– Their company worked hard for an image of social sustainability		X	

Based on the empirical work of Marshall et al.[46]; Nat and Agrawal[47]; Sudusinghe and Seuring[48]

through a level of disruptive sustainable innovation where it changes processes, generates products, raw materials that guarantee the well-being of the people outside the company, as well as the final customer (Table 3).

[46] Marshall, McCarthy, McGrath and Claudy, "Going above and beyond: how sustainability culture and entrepreneurial orientation drive social sustainability supply chain practice adoption", pp. 440–441.
[47] Nath and Agrawal, "Agility and lean practices as antecedents of supply chain social sustainability", International Journal of Operations & Production Management, pp. 1610–1611.
[48] Jayani Ishara, Sudusinghe and Stefan Seuring, "Social sustainability empowering the economic sustainability in the global apparel supply chain", p. 18.

Table 2 Proposed dimension and indicators for the Level 2 construct: Basic SS practices

Dimension and item by level of incorporation of social sustainability		Internal society (human resources)	Immediate external society (community)	Stakeholders (partners, corporate order, suppliers)
Level 2. Basic practices in the supply chain				
1. Internal health and safety	– Monitoring your key supplier's compliance with your health and safety requirements			X
	– Organisational promotion of healthy and economical food	X		
	– Sending health and safety questionnaires to your key supplier to monitor compliance			X
	– Monitored key supplier's commitment and strict policies to health and safety improvement objectives	X		
	– Your supplier guarantees drinking water and sanitation for their employees	X		
	– Introduced employee health and safety audit and compliance systems with key supplier	X		
2. External community benefit	– Our suppliers participate in philanthropic activities		X	
	– Philanthropic activities		X	
	– Actively collaborate to carry out skills development programs for unemployed youth		X	

(continued)

Table 2 (continued)

Dimension and item by level of incorporation of social sustainability		Internal society (human resources)	Immediate external society (community)	Stakeholders (partners, corporate order, suppliers)
	– Disaster/emergency planning or response		X	
3. Suitable working conditions	– Your supplier guarantees proper working conditions	X		
	– Research and development	X		
	– Designed systems for work/family balance throughout the supply chain with key supplier	X		
	– Actions against workplace violence and harassment	X		
	– Freedom of association and collective bargaining	X		
	– Wage and benefit negotiations throughout the supply chain	X		
4. Inclusiveness and gender equity	– Promoting all employees equally on the basis of merit	X		
	– Respect rights and privileges of employees regardless of age, sex, race, community, religion or nationality	X		

(continued)

Table 2 (continued)

Dimension and item by level of incorporation of social sustainability		Internal society (human resources)	Immediate external society (community)	Stakeholders (partners, corporate order, suppliers)
	– Ensure that there is no gender discrimination in our suppliers	X		
	– Suppliers hire locals, women, people with disabilities, the marginalized and minorities		X	
	– Women's participation in leadership	X		
5. Supplier development	– Your supplier helps develop local suppliers (supplier's supplier)			X
	– You have helped your key supplier obtain ISO 4500, SA8000, ISO 26000, ISO 20400 or other management system certification			X
6. Capacity building	– Training for availability of skilled manpower	X		
	– Opportunity for professional development and education	X		
	– Development of soft skills (emotional intelligence, collaborative, motivation, problem-solving)	X		

(continued)

Table 2 (continued)

Dimension and item by level of incorporation of social sustainability		Internal society (human resources)	Immediate external society (community)	Stakeholders (partners, corporate order, suppliers)
7. Ethics and regulatory respect	– You developed an ethical code of conduct system with your key supplier			X
	– Anti-corruption policies			X
	– Ethical behaviour and respect for regulations and legal norms			X
8. Conditions of child labour and trafficking	– Abolition of child labour	X		
	– Elimination of forced labour	X		

Based on the empirical work of Marshall et al.[49]; Mani et al.[50]; Nat and Agrawal[51]; Sudusinghe and Seuring[52]

[49] Marshall, McCarthy, McGrath and Claudy, "Going above and beyond: how sustainability culture and entrepreneurial orientation drive social sustainability supply chain practice adoption", pp. 440–441.

[50] Venkatesh Mani, Angappa Gunasekaran and Catarina Delgado, "Enhancing supply chain performance through supplier social sustainability: An emerging economy perspective", International *Journal of Production Economics* 195, (2018): 259–272.

[51] Nath and Agrawal, "Agility and lean practices as antecedents of supply chain social sustainability". International Journal of Operations & Production Management, pp. 1610–1611.

[52] Jayani Ishara, Sudusinghe and Stefan Seuring "Social sustainability empowering the economic sustainability in the global apparel supply chain", p. 18.

Table 3 Proposed indicators for the Level 3 construct: Advanced SS practices

Dimension and item by level of incorporation of social sustainability	Internal society (human resources)	Immediate external society (community)	Stakeholders (partners, corporate order, suppliers)
Level 3. Advanced practices			
– Your company developed new products/processes with its key supplier that benefited workers throughout the supply chain	X		
– Your company developed new products/processes with its key supplier that reduced employee health and safety risks		X	
– Your company developed new products/processes with its key supplier that reduced consumer health risks		X	
– Your company has changed its supply chain strategy to address demands from non-governmental organisations and community groups in the supply chain		X	
– Your company developed new products/processes with its key supplier that provided fair margins to all suppliers			X
– Actively collaborate to carry out skills development programs for unemployed youth			X
– We guided suppliers in the implementation of occupational health and safety measures	X		
– Your company has changed its supply chain strategy to minimise negative impacts on the communities around its supply chain operations		X	

(continued)

Table 3 (continued)

Dimension and item by level of incorporation of social sustainability	Internal society (human resources)	Immediate external society (community)	Stakeholders (partners, corporate order, suppliers)
– Your company has changed its supply chain strategy to focus on fair trade throughout the supply chain			X
– Your company has changed its supply chain strategy to make social sustainability data (ethical code of conduct/impact on communities) throughout our supply chain publicly available		X	

Based on the empirical work of Marshall et al.[53]; Mani et al.[54]; Nat and Agrawal[55]; Sudusinghe and Seuring[56]

5.3 Integrity of the Essential Aspects of Human Welfare

This proposes a model of three constructs, one for each level of sustainability integration moments: (1) Orientation towards sustainability (eight items, Table 1), (2) Basic practices (31 items in eight dimensions, Table 2) and (3) Advanced practices (ten items, Table 3). The contribution is to make explicit the conceptualisation of the contexts in which social sustainability is assimilated and to point out each item in which context it impacts, whether internally, externally or on the stakeholders.

[53] Marshall, McCarthy, McGrath and Claudy, "Going above and beyond: how sustainability culture and entrepreneurial orientation drive social sustainability supply chain practice adoption", pp. 440–441.

[54] Mani, Gunasekaran and Delgado, "Enhancing supply chain performance through supplier social sustainability: An emerging economy perspective", p. 269.

[55] Nath and Agrawal, "Agility and lean practices as antecedents of supply chain social sustainability", International Journal of Operations & Production Management, p. 1616.

[56] Jayani Ishara, Sudusinghe and Stefan Seuring, "Social sustainability empowering the economic sustainability in the global apparel supply chain", p. 18.

6 Conclusions

The central question that this chapter sought to answer was, What are the possible paths for social sustainability practices in the Mexican space sector supply chain? Following the analysis of the arguments, it can be concluded that Mexico has sought to advance quantitatively and qualitatively in order to approach and benefit this sector. However, there is still a long way to go, as the strategies and actions that have been formulated and implemented seek to improve the conditions of the companies through certifications, investment in technology and attracting international companies, but leave aside important areas such as the environmental and social impacts these activities may have. This is essential to achieve sustainability within the supply chain, i.e., by integrating environmental, social and economic impacts throughout the life cycle of products and processes provided in the aerospace sector, with a long-term vision in which stakeholders promote sustainable development.[57] The Mexican (aero)space sector requires actions and strategies that involve all the stakeholders in the chain so that they participate in achieving these objectives with commitment, responsibility, in an integral and transversal manner.

Efforts to promote space sustainability have evolved since the stakeholders involved have understood that any action, they take can have a long-term impact on future space activities. However, the concept of space sustainability is still very ambiguous, since the idea of implementing a model that guarantees the use of outer space has become widespread, without establishing the sustainable needs of the processes carried out on Earth to achieve the successful development of a space mission, and thus contribute to the fulfilment of the SDGs in a comprehensive manner.

Sustainability should be seen as an integrator of environmental, social and economic systems that requires that the social and ecological dimensions of policies be considered at the same time as the economic, trade, energy, technology and other dimensions. It is important that national and international institutions and programs examine and adopt these actions, that is, that the existing balance between man and the surrounding resources be reflected in a harmony between economic, social and ecological resources. In addition to the above, the Mexican space sector has the global challenge of catching up in terms of socially sustainable supply chain management, which is to pay special attention to the study and implementation of social care in its manufacturing suppliers. It is clear that collaborating with a socially sustainable partner not only fulfils an ethical condition, but also meets a global demand; furthermore, it generates a better social performance in the context of the supply chain.[58] Social sustainability also brings with it, in the medium and

[57] United Nations Global Compact and Business for Social Responsibility, *Sustentabilidad de la cadena de suministro, una guía práctica para la mejora continua,* (New York: United Nations, 2010), p. 5.

[58] Vishnu Nath and Rajat Agrawal, "Agility and lean practices as antecedents of supply chain social sustainability", International Journal of Operations & Production Management, pp. 1610–1611.

long term, advantages in the operational performance of the supply chain.[59] It has even been shown that socially sustainable performance positively affects economic sustainability performance in the supply chain.[60]

Although the contribution of this research is to provide answers for action, it seems to us that the most important contribution are the questions. Proposing elements for introspective reflection on the sector based on the questions indicated. Is the aerospace sector considering the social footprint that its productive and scientific activity leaves along its supply chain? Are the Mexican space sector manufacturers oriented towards social sustainability (do they know what it implies, do they disseminate what it means, do they accept its importance?)? Is there a notion of the social impact that the sector's activity leaves in the operation of its different links in the supply chain, in its internal population, in the social fabric of its external community, in its minor suppliers? Does the Mexican space sector monitor the state of social aspects in its suppliers? What is its situation in terms of education and training of employees, inclusion, gender equity, salary fairness, health and safety, ethics, abolition of child labour, anti-corruption policies?

This chapter proposes a model that can be used as a starting point for future lines of research. In a next methodological step, it is proposed to validate these indicators with experts from the Mexican aerospace sector and, by means of a Delphi technique or a best-worst-method, to prioritise, rank and rule out items and achieve measures relevant to the country's supply chain. Subsequently, the model must be contrasted with reality, initially with the adjustments of a qualitative case study to verify that the indicators are supported when filtered with the reality of the companies themselves. After the contrast with experts and cases, the model would be ready to be subjected to an empirical statistical sample study and to verify the generalisation capacity of the dimensions and indicators.

Lisette Farah Simón Ph.D., Completed her undergraduate studies at the Faculty of Engineering at UNAM (National Autonomous University of Mexico) where she graduated as an Industrial Engineer; she obtained a master's degree in Mechanical Engineering from the same faculty and received the degree of Doctor of Administration Science, with honours, from the FCA (Faculty of Accounting and Administration-UNAM). She has participated in several R&D projects, among which her participation stands out: in the Macro Project "Administration and Sustainability" of the FCA with the project management of technology with a sustainable perspective in the space industry in Mexico; in the Space Science and Technology Network (Thematic Network) and in the University Space Program, with a project on space sustainability; as well as the research she carried out on the design of an adjustable socket for lower limb prostheses; which has been protected through an invention patent. She is a member of the National System of Researchers of Conacyt (National Council of Science and Technology).

[59] Mani, Gunasekaran and Delgado, "Enhancing supply chain performance through supplier social sustainability: An emerging economy perspective", p. 269.

[60] Jayani Ishara, Sudusinghe and Stefan Seuring, "Social sustainability empowering the economic sustainability in the global apparel supply chain", p. 18.

Miguel Angel Reyna-Castillo Philosopher, Master's in Human Development and Doctorate in Strategic Business Management. Post-doctorate in Corporate Social Sustainability, National Council of Science and Technology-Conacyt (in progress). Post-doctorate in Sustainable Management, Research Division of the Faculty of Accounting and Administration of the UNAM (National Autonomous University of Mexico), DIFCA, Research Division of the Faculty of Accounting and Administration (Doctoral research placement at the Faculty of Administration of the Universidad de Medellín. Member of the National System of Researchers (SNI). Undergraduate and graduate research professor. Pedagogue in research methodology: protocol generation, bibliometric review, theory building, empirical case method and statistics. Lines of research: Strategic management, gender in organisations, management theories and social sustainability.

Hugo Javier Buenrostro-Aguilar Ph.D., Degree in International Relations (Faculty of Political and Social Sciences FCPyS-UNAM); Master's in international business administration (Faculty of Accounting and Administration, FCA-UNAM); Doctorate in Administration Science (Faculty of Accounting and Administration, FCA-UNAM). Full-time professor attached to the Research Division of the Faculty of Accounting and Administration (FCA) of the UNAM (National Autonomous University of Mexico). Lecturer and guest speaker by the Centre for Social Studies and Public Opinion, Chamber of Deputies; the Strategic Research Institute of the Mexican Navy and the Centre for Naval Superior Studies, Mexican Navy Secretariat; and, by the Superior School of War of the Secretariat of National Defense. Member of the Mexican Association of International Studies; the Administration and Sustainability Macroproject of the FCA-UNAM; and, from the University Seminar on Social Entrepreneurship, Sustainable Administration and Comprehensive Training at Middle and Higher Levels at the Universidad Nacional Autónoma de México (SUESA).

Bolivian CanSat Contest: Promoting Space Science and Technology

Soliz Jorge, Puma-Guzman Rosalyn,
and César Andrés Cabrera Cesar

Abstract

Attracting young people's interest into science has always been a challenge and something that is rarely achieved. This is mainly due to the misconception that science is all about numbers and calculations, which is obtained by the difficulty presented in their years in school. Although it is true that to understand an exact natural science it is necessary to have a basic knowledge in areas related to physics, mathematics, or chemistry. Many of these subjects are often focused in a non-applicational way and for this reason most students don't understand in a clear way the main concepts of these subjects. Nowadays, this is slowly changing due to the growing number of schools, associations and universities that are showing more interest in prioritizing the promotion of natural scientific areas of study and investigation. In South America the most common way to promote this interest in natural science and its applications is through science fairs, and contests on different topics (robotics, astronomy, programming, video games, etc.). In the region, the best results were seen in contests, which were initially developed in universities but were later implemented in schools as well. Regarding the space area there are different types of contests, but the one that attracts attention and stands out most is the CanSat Contest that has been implemented more and more throughout different countries in South America. A striking feature of this contest is its slogan "You can build a satellite". The

S. Jorge (✉)
Researcher Radiocommunications Lab (LRC), Universidad Privada Boliviana, Cochabamba, Bolivia
e-mail: jorgesoliz@upb.edu

P.-G. Rosalyn
Industrial Engineering and Systems, Universidad Privada Boliviana, Cochabamba, Bolivia

C. A. Cabrera Cesar
Exact Science Department, Universidad Privada Boliviana, Cochabamba, Bolivia
e-mail: cesarcabrera@upb.edu

A. Froehlich (ed.), *Space Fostering Latin American Societies*, Southern Space Studies,
https://doi.org/10.1007/978-3-031-20675-7_5

aim is to involve students in space projects by providing a brief yet complete knowledge about the design and implementation of a space mission. This contest focusses on applied science but also boosts the participants to gain more confidence to achieve greater goals in the space field. The principal objective of the CanSat Contest apart from simulating a real space mission is to involve participants in the design and construction processes of components that make up the different subsystems of a satellite. Like for example the electrical power system, communication system, payload, etc. All these sub-systems must enter in a container in the size of a soda can which will be positioned at an altitude between 300 to 500 m above the surface and have to transmit a signal (wireless, not Wi-Fi) in real-time to a ground station connected to a computer, where the data and results obtained by the satellite are displayed. The CanSat Contest reflected the best-selected option by the company Sur Aerospace and the EMI University to foment science and technology space groups in Bolivia. Groups that were formed by students from different universities were offered the possibility to compete in this type of contest that takes place in many countries around the world and is qualified by the same rules. For this reason, these institutions held the first CanSat contest in Bolivia, in which involved eight participating groups and more than 40 students.

1 Introduction

Nowadays, the amount of young people that decide to study careers related to natural science, technology, engineering, and mathematics (STEM) is decreasing. This trend motivates many private and government entities to develop programs designed to change the direction of this tendency.

Professor Bob Twiggs of Stanford University was the one who proposed the idea of building and launching three picosatellites for scientific research.[1] Each of these with a volume of 350 cm^2, a mass of 500 g, a size structure like a can of soda and a capacity to carry 1,8 kg and designed to be sent to space within a rocket. This idea was proposed in the first "University Space Systems Symposium" held in Hawaii in 1998 and led to a project that began in 1990, called "A rocket launch international student satellites (ARLISS)", conformed by mostly American and Japanese participants. At last, it ended with the launch in September of the same year.

A CanSat contest is a real satellite simulation, integrated within the volume and shape of a soda can. CanSat aims to reproduce at scale the process by which

[1] Roger Walker, Technology cubesat Manager (ESA), https://www.esa.int/Space_in_Member_States/Spain, accessed 29 August 2022.

a satellite is designed, built, tested, launched and operated. The challenge for students is to fit all the major subsystems found in a satellite, such as power supply, sensors, and a communication system, into this reduced volume. The CanSat is designed to be launched either by a rocket, from a platform, drone, or captive balloon to an altitude of around 500 m. Once this is achieved, its mission begins. All in all, this process is classified as a scientific experiment, in which the main objective is to achieve a safe landing and to analyze the collected data.[2]

1.1 Promotion of Space Science

The worldwide main entities promoting space science, The National Aeronautics and Space Administration (NASA) and the European Space Agency (ESA), exhaust all possible means of communication, such as Facebook, Instagram, Twitter, among others, YouTube channels, applications for mobile devices, face-to-face and virtual conferences, contests, podcasts, etc.

At NASA, the STEM engagement portfolio consists of a diverse set of opportunities, activities, and products, encompassing internships; scholarships; student learning opportunities (challenges, competitions, and other experiences); informal education and non-formal learning activities; educational products, tools and platforms; support to the educator; competitive awards to educational research and development institutions, and institutional support.[3]

On the other hand, ESA has the European Space Educational Resources Office (ESERO) program which is based in ESA's 20 official member States and designed for young European students. In order for students to feel more comfortable and familiar with the sciences in general, ESERO uses spatial context to make the teaching and learning of STEM subjects more engaging and accessible. Additionally, ESERO activities promote the availability of STEM subjects and learning to all students, demolishing the misconception that science is only for geniuses, talented and/or "smart" people.[4]

1.2 Contests in North America and Europe

One of the mostly used resources to motivate and train future generations of researchers are competitions. NASA's "Commitment STEM" offers a wide variety of competitions, carrying this purpose. Within the space sciences area, the

[2] Ramon Carrasco Duboué, Samuel Vásquez Hernández "Nanosatélite basado en microcontroladores PIC: CanSat", III Congreso Virtual, Microcontroladores y sus Aplicaciones – Instituto de Geofísica y astronomía, Cuba (2014).

[3] Brian Dunbar, National Aeronautics and Space Administration (NASA), 10 May 2022, https://www.nasa.gov/stem/about.html, accessed 19 July 2022.

[4] The European Space Agency (ESA), July 2022, https://www.esa.int/Education/Teachers_Corner/European_Space_Education_Resource_Office, accessed 19 July 2022.

main objectives of these competitions are to inspire, engage, educate and employ. Among these, one can mention many representative examples. Firstly, "DEVELOP" is a competition that provides the opportunity to investigate and address both environmental and political concerns through the practical application of NASA's Earth science information and geospatial data. Then, the "AstroPhoto Challenges" competition provides the opportunity to use real astronomical data and tools to create one's own images from the Eta Carinae star system. Thirdly, the "FLOATing DRAGON Balloon Challenge" is designed to de carried out by several teams and their faculty advisors, who are in charge of designing and prototyping ideas for a guided data vault retrieval system. Also, in "Short film competition", all film entries should be no longer than 10 min and at least 10% of each should use NASA's archival footage to create unique NASA-inspired masterpieces.

In ESA there is the "LandsatCraft" competition and many others promoting space interest. In this sense, ESERO also has a series of activities and competitions such as "Astro Pi Challenge", which demand participants to code a computer program on the Raspberry Pi computer that is on board the International Space Station. On the other hand, "Mission X", there is a competition that focuses on the physical and nutritional condition of the participants, allowing participants to simulate the practice of astronaut's training. Then, on the "Moon camp challenge", with the use of 3D modeling tools participants have to design a lunar settlement. Environmentally focused, the "Climate detectives" competition demands the identification of a problem related to the climate and the design of a way to monitor it. And last but not least, there is a project that tries to simulate a space mission called CanSat.[5]

Likewise, the American Astronautical Society (AAS) also organizes an annual design, construction and launch competition for students with space-related topics. Its main goal is to give students the satisfaction of being involved with the end-to-end life cycle of a complex engineering project, from conceptual design, following with integration and testing, the actual operation of the system and concluding with a summary and post-mission report. Within this context, the CanSat competition meets these needs.[6]

1.3 Contest Latam

Since many countries have already been able to develop their own CanSat competitions, there have already been different educational experiences in recent years. Besides, these competitions were mainly developed by universities, research centers and space agencies.

---

[5] Brian Dunbar, National Aeronautics and Space Administration (NASA), 10 May 2022, https://www.nasa.gov/stem/highereducation/index.html, accessed 20 July 2022.

[6] The American Astronautical Society (AAS), 12 June 2022, https://cansatcompetition.com/, accessed 20 July 2022.

In Latin America, this event has begun to be officially organized by Mexico as the pioneer. This country began working in this area in 2013, with the successful launch of a CanSat, which was the product of a course-workshop organized by the Program for Aerospace Experimentation and Research (PROEXIA) of Pozo Rica.[7] This project was followed by the University Space Program (PEU) of the National Autonomous University of Mexico (UNAM). PEU has been organizing this event since 2014. At first it started only for UNAM students, but in its latest versions it was nationally launched. In this competition, the team that obtained the first place was selected as the representative to participate in the annual CanSat competition. Also, the University Center for Exact Sciences and Engineering (CUCEI) of the University of Guadalajara (UDG) has been organizing the event since 2014. In the first versions, weather balloons were used to launch the CanSat at more than 4.000 m above sea level, currently drones are used for launches under an operations plan coordinated with the Local Command of the General Directorate of Civil Aviation (DGAC).

In Argentina, the activities with CanSats began in 2004 in schools and, which were supported and organized by the Association of Experimental Rocketry and Modeler (ACEMA) of Argentina as well as the Argentine Association of Space Technology (AATE). Currently, these contests are also an initiative organized by the Ministry of Science, Technology and Innovation (MINCyT) and the National Commission for Space Activities (CONAE). Its major aim is to awaken in the Argentine youth population the spirit of scientific research. The call has as a target audience of secondary students and they must form groups and go through the following stages:

- Mandatory virtual training.
- Writing and presentation of the project.
- Selection of projects to be specified and sending of the material kit.
- Construction, validation and monitoring of CanSats.
- Launch campaign.

Bolivia has joined this trend by organizing of the first CanSat contest in 2018. It was developed by the Sur Aerospace company with the support of different academic institutions.[8]

[7] Wikipedia, https://es.wikipedia.org/wiki/Cansatialibre, accessed 29 August 2022.

[8] Arruabarrena-Mariana, Fernández-Agustín, Medel-Ricardo and Mori-Luciano, "Estudio Bibliográfico del Estado del Arte del Desarrollo y Aplicaciones Educativas de CanSats", X Congreso Argentino de Tecnología Espacial, 10 to 12 April 2019, Buenos Aires, Argentina.

2 Promoting Space Science in Bolivia

2.1 Background

Like many countries in South America, in Bolivia the development of the aerospace technology area is very scarce. While aerospace engineering in the world is constantly changing due to new technologies and research.

Bolivia began its journey in the space area on 10 February 2010, with the creation of the ABE ("The Bolivian Space Agency"), whose main objective is to manage and execute the implementation of the Tupac Katari Satellite Project. Its specific functions are[9]:

- Manage and execute the implementation of the Tupac Katari Satellite Project.
- Contribute to the reduction of the digital gap in the country with space technologies.
- Manage and execute space projects, remote sensing, telecommunications and information and communication technologies.
- Provide space services, remote sensing, telecommunications and information and communication technologies.
- Promote and execute technology transfer and training of human resources in space technology.
- Promote and execute the implementation of satellite applications for use in social, productive, defense, environmental and other programs.

Besides many universities and groups of students made some efforts to increase the interest in this area by space dissemination talks and space-themed congresses.

2.2 CanSat Contest

The CanSat competition seeks to simulate a real space mission. In this way, apart from the resulting spacecraft, within a space mission, there are many aspects to consider.

A space mission is constructed beginning with the initial idea of the project and ending with the analysis of results obtained from the spaceship tool. Due to this, in a CanSat competition, all the parts that make up the relevant individual space mission are evaluated and taken into account for the final scoring.

2.2.1 Real Space Project and Phasing

A CanSat contest offers a unique opportunity for students from different universities to have a first practical experience of a real space project. They are responsible

[9] Luis Alberto Arce Catacora, Decreto Supremo No. 4735, 8 June 2022, https://www.lexivox.org/norms/BO-DS-N4735.xhtml, accessed 23 July 2022.

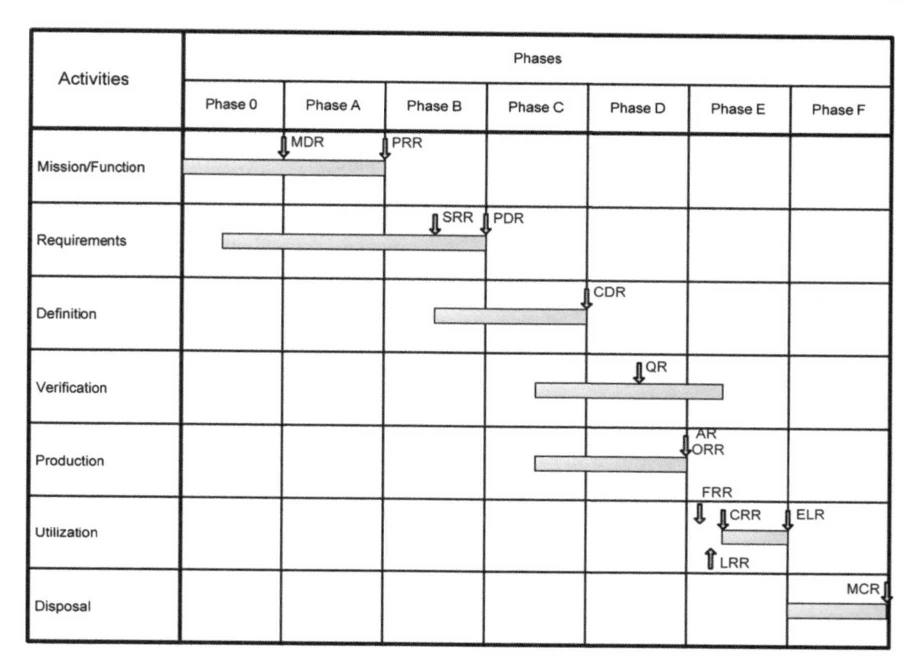

Fig. 1 Typical project life cycle according to the ECSS standard[10]

for all aspects: selecting mission objectives, designing the CanSat, integrating the components, testing, preparing the launch, and then analyzing the data.

According to the ECSS standard (European Cooperation for the Standardization of Space) a space mission is divided into seven phases with their respective revisions from its approach to its disuse (see Fig. 1).

According to common standards, a space mission is made up of the following revisions (Table 1).

2.2.2 Standards and Phases

The ECSS established in 1993, is an initiative established to develop a coherent and unique set of user-friendly standards for all European space activities.

The goal of building such a standardization system, at the European level, is to minimize the life cycle cost, while continuously improving the quality, functional integrity, and compatibility of all elements in a space project. This goal is achieved by applying common standards for project management and hardware and software development and testing.

[10] R. Puma-Guzman, J. Soliz, "Launch Management of a Nanosatellite for Bolivia", Space Fostering Latin American Societies Developing the Latin American Continent Through Space Part 2, Springer, 2021.

Table 1 Reviews of mission[11]

Acronymous	Definition	Main objectives
MDR	Mission Definition Review	• Release the mission statement • Assess the preliminary technical requirements specification and programmatic aspects
PRR	Preliminary Requirements Review	• Release of preliminary management, engineering and product assurance plans • Release of the technical requirements specification • Selection of system and operations concept(s) and technical solutions, including model philosophy and verification approach, to be carried forward into Phase B
SRR	System Requirement Review	• Release of updated technical requirements specifications • Assessment of the preliminary design definition • Assessment of the preliminary verification program
PDR	Preliminary Design Review	• Verification of the preliminary design of the selected concept and technical solutions against project and system requirements • Release of final management, engineering and product assurance plans
CDR	Critical Design Review	• Assess the qualification and validation status of the critical processes and their readiness for deployment for Phase D • Confirm compatibility with external interfaces • Release the final design • Release assembly, integration and test planning
QR	Qualification Review	• To confirm that the verification process has demonstrated that the design, including margins, meets the applicable requirements
AR	Acceptance Review	• To verify that all deliverable products are available per the approved deliverable items list • To authorize delivery of the product • To release the certificate of acceptance
ORR	Operation Readiness Review	• To verify readiness of the operational procedures and their compatibility with the flight system

(continued)

[11] European Cooperation for Space Standardization (ECSS), "Space Project Management", 6 March 2009.

Table 1 (continued)

Acronymous	Definition	Main objectives
FRR	Flight Readiness Review	• Verify that the flight and ground segments including all supporting systems such as tracking systems, communication systems and safety systems are ready for launch
LRR	Launch Readiness Review	• Declare readiness for launch of the launch vehicle, the space and ground segments including all supporting systems such as tracking systems, communication systems and safety systems • Provide the authorization to proceed for launch
CRR	Commissioning Result Review	• The commissioning result review is held at the end of the commissioning as part of the in-orbit stage verification. It allows declaring readiness for routine operations/utilization
ELR	End of Life Review	• To verify that the mission has completed its useful operation or service

According to these standards, a space mission consists of the following seven phases[12]:

Phase 0: Mission Analysis and Needs Identification.

- Support the customer by identifying their needs.
- Propose possible concepts of the system.
- Document mission objectives and needs.

Phase A: Feasibility.

- Refine and finalize the definition of mission needs.
- Specification of high-level requirements.
- Baseline of requirements.

Phase B: Preliminary Definition.

- Operational Requirements.
- Preliminary subsystem requirements.

[12] European Cooperation for Space Standardization (ECSS), "Space Project Management", 6 March 2009.

- Specification of technical requirements.
- Requirements for the management of the project.
- Preliminary design of subsystems.

Phase C: Detailed Definition.

- Detailed definition of the architecture.
- Defining processing tasks.
- Verification of designs against requirements.
- Verification of designs against technical specification.

Phase D: Qualification and Production.

- Finalize system development.
- Component Assembly.
- Planning and execution of quality tests.
- Integration test plan.
- Integration Tests.
- Acceptance.
- Tests.

Phase E: Utilization.

- Operations Plan.
- Ensures the project is on schedule.
- Ensures availability of deliverables.
- Product ready for launch.

Phase F: Disposition.

- Evaluation of flight activities after recovery.
- Identification of anomalies.
- Anomaly and Error Mitigation Plan.
- Lessons learned.

2.2.3 Competition Project

The Bolivian company Sur Aerospace and the EMI university organized in 2018 the "1st. Bolivia's CanSat Contest". In this the first edition, eight universities from different departments of Bolivia participated. The Sur Aerospace company is the first private initiative in the country, it that was born with the purpose of addressing the aerospace field. Among its main objectives are the dissemination of science, the strengthening of aerospace knowledge, the motivation of Bolivia's youth population (see Figs. 2 and 3).

Fig. 2 Members of the Bolivian CanSat Contest Company Sur Aerospace in 2018

Fig. 3 CanSat of Universidad Privada Boliviana (UPB) in 2018

Fig. 4 Hacklab Team, "Winner of CanSat Bolivia 2018"

2.2.4 General Objective of CanSat Contest

Involve the university's community in the field of aerospace engineering, through the manufacture of a ground satellite model called CanSat.

Specific objectives of the Bolivian CanSat contest in 2018 were:

1. Generate creative and innovative ideas using space technology in the solution of a mission for an educational picosatellite CanSat.
2. Promote collaborative work using aerospace science and technology.
3. Venture into scientific research on topics related to space.
4. Design and construction of components of the subsystems of a satellite for terrestrial use.
5. Integrate multidisciplinary groups in the development of CanSat subsystems.

2.2.5 Primary and Secondary CanSat Missions

Primary Mission

Teams had to design, build, and program a CanSat to accomplish the following compulsory primary mission: After the release and during descent, the CanSat measures the following parameters and transmit the following data as telemetry to the ground station at least once every second: the air temperature and air pressure.

The teams analyzed the data obtained and showed them in graphs: altitude versus time and temperature versus altitude; in a post-flight analysis.

Secondary Mission

Students had to develop a side quest of their choice. They may be inspired by other real satellite missions, a need for scientific data for a specific project, a technology demonstration for student-designed components, or another mission that CanSat can accommodate. Some examples of missions are listed below, but teams are free to design a mission of their choice, if they can demonstrate to have some scientific, technological or innovative value. Teams should also keep in mind the limitations of the CanSat mission profile and focus on the feasibility (both technical and administrative) of their chosen mission.

- Photography
- Photogrammetry
- Real-time Video Streaming
- CanSat Orientation
- CanSat Position
- Object Recognition and Tracking
- Air temperature and Pressure
- Advanced Telemetry
- Telecommand
- Targeted Landing
- Landing System and Planetary Probe.

2.2.6 Requirements and Restrictions

The CanSat hardware and mission must be designed following these requirements and constraints[13]:

1. All the components of the CanSat must fit inside a standard soda can (115 mm height and 66 mm diameter), except for the parachute. Radio antennas and GPS antennas can be mounted externally (on the top or bottom of the can, but not on the sides), depending on the design.
2. The antennas, transducers, and other elements of the CanSat cannot extend beyond the can's diameter until it has left the launch vehicle.
3. The mass of the CanSat must be between 300 and 350 g. CanSats that are lighter must take additional ballast with them to reach the 300 g minimum mass limit required.
4. Explosives, detonators, pyrotechnics, and inflammable or dangerous materials are strictly forbidden. All materials used must be safe for the personnel, the equipment, and the environment.

[13] The European Space Agency (ESA), "European CanSat Competition Guidelines", 2018.

5. The CanSat must be powered by a battery and/or solar panels. It must be possible for the systems to be switched on for four continuous hours.
6. The battery must be easily accessible in case it has to be replaced/recharged.
7. The CanSat must have an easily accessible master power switch.
8. Inclusion of a retrieval system (beeper, radio beacon, GPS, etc.) is recommended.
9. The CanSat should have a recovery system, such as a parachute, capable of being reused after launch. It is recommended to use bright colored fabric, which will facilitate recovery of the CanSat after landing.
10. The parachute connection must be able to withstand up to 1.000 N of force. The strength of the parachute must be tested to ensure that the system will operate nominally.
11. For recovery reasons, a maximum flight time of 120 s is recommended. If attempting a directed landing, then a maximum of 170 s flight time is recommended.
12. A descent rate between 8 and 11 m/s is recommended for recovery reasons. In case of a directed landing, a lower descent rate of 6 m/s is recommended.
13. The CanSat must be able to withstand an acceleration of up to 20 G.
14. The CanSat must be flight-ready upon arrival at the launch campaign. A final technical inspection of the CanSats will be done by authorized personnel before launch.

2.2.7 Contest Evaluation

The teams been evaluated continuously, considering the following aspects[14]:

Educational Value

The jury will consider the content and quality of the Pre-launch Report and team presentations, the level of effort made by the team, and how much the team appears to have learned throughout the project.

Technical Achievement

Innovative aspects of the project will be judged (e.g. the mission selected and the hardware/software used). The jury will also take into account how the teams obtained the results, how reliable and robust the CanSat was, and how the CanSat performed. If the CanSat did not succeed in accomplishing its mission, but the team is able to explain the reasons why and suggest improvements, it will be taken into account positively.

[14] Ibid.

Table 2 Participants in CanSat contest

State/town	University/Team	Number of participants
Cochabamba	Escuela Militar de Ingeniería (EMI)	5
	Universidad Privada Bolivia (UPB)	7
	Hacklab Team	3
Santa Cruz (Riberalta)	Escuela Militar de Ingeniería (EMI)	4
Tarija	Elite-Sat Team	2
Santa Cruz	Escuela Militar de Ingeniería (EMI)	4
Potosí	Universidad Nacional Siglo XX	6
La Paz	Escuela Militar de Ingeniería (EMI)	4

Teamwork
The Jury will assess how well the team worked together on the assignment, the distribution of tasks, the planning and execution of the project, and the team's success in obtaining the necessary funding, support, and advice.

Outreach
The team will be awarded points on how well the project was communicated to the school and the local community, taking into account web pages, blogs, presentations, promotional material, media coverage, etc.

2.3 Experiences

The CanSat Bolivia 2018 Contest had the participation of eight teams from the different universities of the country, this competition was held at the Quillacollo Model Aircraft Association on 29 September 2018 between 8:00 a.m. and 1:00 p.m. The juries of the contest were Msc. Erick Pozo, Msc. Ricardo Colpari and Ph.D. Jorge Soliz.[15]

The projects presented had to fulfill the missions specified in Sect. 2.2.5. The groups that participated were (Table 2).

After having carried out the evaluation of each of the projects as mentioned in Sect. 2.2.7, the results were:

First Place and Winner Team: Hacklab Team
Participants: Students from the Universidad Mayor de San Simón and Universidad Simón I. Patiño.

Experiments: Measurement of environmental variables (humidity and temperature), Aerial shots of the terrain through a camera, GPS Sensor Readings

[15] Catherine Camacho, Los Tiempos, 30 September 2018, https://www.lostiempos.com/doble-click/vida/20180930/hacklab-upb-emi-ganan-concurso-picosatelites, accessed 29 August 2022.

and Measurements with (Accelerometer, Magnetometer, Barometer, Gyrometer) (Fig. 4).

Marking scheme	
Educational value (20%)	15%
Technical achievement (50%)	46%
Teamwork (15%)	15%
Outreach (15%)	15%
Total %	91%

Second Place: UPB Team
Participants: Students of the Universidad Privada Boliviana.

Experiments: GPS sensor readings, atmospheric probe, and speed measurements.

Marking scheme	
Educational value (20%)	15%
Technical achievement (50%)	38%
Teamwork (15%)	15%
Outreach (15%)	15%
Total %	83%

Third Place: EMI-Cochabamba Team
Participants: Students of the Military School of Engineering.

Experiments: Measurement of environmental variables (humidity, temperature, UV rays), GPS Sensor readings.

Marking scheme	
Educational value (20%)	10%
Technical achievement (50%)	33%
Teamwork (15%)	15%
Outreach (15%)	15%
Total %	73%

3 Conclusions and Recommendations

The first CanSat contest in Bolivia provided the opportunity for university students to exchange knowledge and experiences with other people who have the same passion for space science and technology. The contest promoted interest and motivation in the participants to continue studying and specializing in space science and technology. The contest was followed by local communication media such as television and newspapers and revealed the processes that young people are capable of developing. The contest had a very good reception in its first version, unfortunately due to political aspects and the 2020 pandemic, it was not held in the years to follow, but it is expected that the competition will be held again next year.

It should be noted that many groups and teams had problems in submitting reviews and technical reports of the different phases of the mission because they were not used to this way of working, something that is fundamental to understand a space mission. Due to this, the contest helped the participants to acknowledge that a space mission is not only about the final result of the mission. The space mission involves many phases of project analysis, development, testing and finally the result of the mission.

***

Soliz Jorge is a Mechanical Engineer with a Ph.D. degree in "Aerospace Science and Technology," and a MSc degree in "Aerospace Engineering" from Universitat Politecnica de Catalunya, España. He worked in projects such as Galactic Suite (space hotel), on space mission analysis; UPCSAT 1, first satellite of the Universitat Politecnica de Catalunya (picosatellite, cubesat 1U); SSETI, "Student Space Exploration and Technology Initiative", on design and construction of satellites (sponsored by European Space Agency), as well as several research projects in Astrodynamics, and design of nano- and picosatellites. Currently, he is professor and researcher at Universidad Privada Boliviana (UPB) (Radiocommunications Laboratory).

Puma-Guzman Rosalyn is an Industrial Engineering and Systems at "Universidad Privada Boliviana" (UPB), Bolivia. She worked in Sur Aerospace (space and aeronautics projects company in Bolivia) on the event organization and project management section. She has experience in space outreach events. She worked in the organization of the first CanSat Bolivia Contest and in CubeSat satellite projects for different universities. She is currently working on her final degree project at UPB, concerning space systems, focused on satellite subsystems and the management of space projects.

César Andrés Cabrera Cesar is a Mathematical Engineer from the Universidad Mayor de San Simón (UMSS) in Bolivia. He is currently in the process of obtaining a master's degree in Economic Engineering from the UMSS and is a professor at the Universidad Privada Bolivia (UPB) teaching mathematics to students of engineering and business sciences. He works in calibration and validation of mathematical models in different areas; stochastic differential equations and dynamic programming.

Dynamic Computational Analysis of a Cubesat Structure to Test a New Material for a Space-Radiation Protection Shield

Bárbara Bermúdez-Reyes, Jorge Enrique Herrera Arroyave, Patricia Zambrano Robledo, Rafael Vargas-Bernal, and Jorge Alfredo Ferrer Pérez

Abstract

This study presents the dynamic structural analysis of a CubeSat structure that comprises three pieces of 6061T6 aluminum (one upper cover, one lower cover, and a body), six integrated shafts, and Seeger rings (without a screw fix), weighing 256,6 g overall. This structure was designed to include a new type of material for protecting the payload and other subsystems within a spacecraft against space radiation. The structural dynamics were simulated under loads induced by the launch vehicle (Ariane V5) using modal and harmonic responses, random vibration, and a response spectrum analysis. A safety factor of 15 and a marginal safety factor of 14 were obtained. The most significant deformation was determined as 0,00357 mm, which indicated that the structure would support the induced loads. The harmonic response analysis found the contributing factors and modal modes in the X-, Y-, and Z-axes of the CubeSat structure. Moreover, frequencies that affected the CubeSat structure without

B. Bermúdez-Reyes
Facultad de Ingeniería (FI UNAM)-Laboratorio de Nacional de Ingeniería Espacial y Automotriz (LN-INGEA) , UNAM-UANL, Juriquilla, Santiago de Querétaro, Qro., México

J. E. Herrera Arroyave
Faculty of Basic Sciences and Engineering, Catholic University of Pereira, Pereira, Colombia

P. Zambrano Robledo
Facultad de Ingeniería Mecánica y Eléctrica, Universidad Autónoma de Nuevo León, San Nicolás de los Garza, México

R. Vargas-Bernal
Instituto Tecnológico Superior de Irapuato, Irapuato, Guanajuato, México

J. A. Ferrer Pérez (✉)
Advanced Technology Unit, School of Engineering, UNAM, Juriquilla, Santiago de Querétaro, Mexico
e-mail: ferrerp@unam.mx

A. Froehlich (ed.), *Space Fostering Latin American Societies*, Southern Space Studies,
https://doi.org/10.1007/978-3-031-20675-7_6

exceeding the natural frequency of vibration (979 Hz) were detected. Using this value, it was possible to simulate the harmonic response analysis at a bandwidth of 2.000 Hz, which did not affect the primary structure. In the random vibration analysis, the energy density spectrum was obtained in a range of frequencies from 30 to 2.000 Hz alongside the validation and computational verification of the independence of the mesh and mathematical model. Based on the obtained results, the CubeSat structure complies with the mechanical–spatial standards and requirements.

1 Introduction

In reality, and based on new exploratory space projects, lighter, small exploration spacecraft structures must be designed using novel techniques to reduce overall mass and enable their use as part of deep space studies and exploration.[1,2] Researchers from different institutions have developed nanosatellite structural systems aimed at reducing the overall mass to allow for adding additional electronic components and to protect payloads from the space environment, with some even proposing novel manufacturing processes (see Footnote 2).

A nanosatellite is a small satellite that can be used for research and the development of accessible space technology. These satellites are characterized by having a mass of 1,33 kg and a $10 \times 10 \times 10$ cm cubic structure. Accordingly, they are known as CubeSats.[3]

Typically, the structural design of a spacecraft depends on the purpose of the specific mission or space mission analysis and design (see Footnote 2). To create the metallic structure of a CubeSat, a dynamic structural analysis methodology must be applied. Such an approach must adjust the structural requirements and restrictions to CubeSat standards.[4] To achieve this, the CubeSat structure must be able to resist static and sinusoidal loads, noise generation during flight, and shock

[1] NASA (2020), NASA Technology Taxonomy, 6 July 2022, https://www.nasa.gov/offices/oct/taxonomy/index.html.

[2] NASA (2020), NASA Explore NASA Small Spacecraft, Strategic Plan, 6 July 2022, https://www.nasa.gov/smallsatinstitute/resources.

[3] Rego A., Pereira M., Greco M. and Peiro E., Design of an 1U CubeSat Plataform Educational Purposes. Master's Thesis, Universida de Federal de Minas Gerais, 2016.

[4] CalPoly CubeSat Design Specification. Technical report, California Polytechnic State University, 2020.

vibrations generated by the separation of the rocket's parts during different launch stages.[5,6]

The CubeSat's attached and fixed parts are created based on the structural design. These parts may include screws, rivets, and spin washers, soldering, and adhesives.[7,8,9,10] When selecting these parts, the following must be considered: (1) structure type; (2) materials; (3) geometry; (4) stiffness and fatigue resistance; (5) joints; (6) payload access; (7) load distribution; (8) preload; (9) disassembly.[11,12]

CubeSat structures have been designed to fit a specific mission or requirement. This was also true in the case of CubeSat named STEP, for which a structure was designed according to the objectives of each subsystem. The characteristics and weight of the components, launching details, deployable subsystems and separation, the configuration of the payload, and available materials were considered in this instance.[13] For the Surya Satellite-1 CubeSat, its structure was designed according to the weight of the complete system, the materials used, manufacturing standards, and the conditions of the launch provider.[14]

---

[5] Raviprasad S. and Nayak N., Dynamic and verification of structurally optimized nano-satellite systems. Journal of Aerospace Science and Technology 1: 78–90, https://doi.org/10.17265/2332-8258/2015.02.005, 2015.

[6] NASA (2020), NASA State of the Art Small Spacecraft Technology, 6 July 2022, https://www.nasa.gov/smallsat-institute/sst-soa.

[7] Ampatzoglou A. and Kostopoulos V. Design, analysis, optimization, manufacturing, and testing of a 2u cubesat. Int. J. Aerospace Eng. 2018. https://doi.org/10.1155/2018/9724263, 2018.

[8] Mathurin F., Guillot J., Stéphan P. and Daidié A., 3d finite element modeling of an assembly process with thread forming screw. J Manuf Sci E 131(4). https://doi.org/10.1115/1.3160377, 2009.

[9] Lennon R., Pedreschi R. and Sinha B. Comparative study of some mechanical connections in cold formed steel. Constr. Build. Mater. 13(3): 109–116, https://www.sciencedirect.com/science/article/pii/S0950061899000185, https://doi.org/10.1016/S0950-0618(99)00018-5, 1999.

[10] Rasmus M., Mäntyjärvi K. and Karjalainen J., Small batch laser welding using light fasteners and laser tack welding, in: Sheet Metal 2011, Key Engineering Materials, volume 473, Trans Tech Publications Ltd, pp. 267–272, https://doi.org/10.4028/www.scientific.net/KEM.473.267, 2011.

[11] Barnes T. and Pashby I., Joining techniques for aluminium spaceframes used in automobiles: Part II—adhesive bonding and mechanical fasteners, J. Mater. Process. Tech. 99(1):72–79. https://www.sciencedirect.com/science/article/pii/S0924013699003611, https://doi.org/10.1016/S0924-0136(99)00361-1, 2000.

[12] Gray P., O'Higgins R. and McCarthy C., Effects of laminate thickness, tapering and missing fasteners on the mechanical behaviour of single-lap, multi-bolt, countersunk composite joints, Compos. Struct. 107:219–230. https://www.sciencedirect.com/science/article/pii/S0263822313003401, https://doi.org/10.1016/j.compstruct.2013.07.017, 2014.

[13] Sadowski T. and Golewski P. Numerical study of the pre-stressed connectors and their distribution on the strength of a single lap, a double lap and hybrid joints subjected to uniaxial tensile test, Arch. Metall. Mater. 52(2). https://doi.org/10.2478/amm-2013-0041, 2013.

[14] Oh H., Jeon S. and Kwon S., Structural design and analysis of 1u standardized step cube lab for on-orbit verification of fundamental space technologies, International Journal of Materials, Mechanics and Manufacturing 2(3): 239–244, https://doi.org/10.7763/IJMMM.2014.V2.135, 2014.

According to Cihan et al.[15] a nanosatellite's structural design must be simulated and evaluated based on mechanical resistance by the finite element method to obtain the structural response according to the launch platform induced loads.[16] One approach for establishing how a CubeSat will behave structurally before manufacturing it is by simulating the dynamic stresses to which it will be exposed (see Footnote 16). Al-Hammadi et al. indicated that it is necessary to determine the four vibrational modes (quasi-static, random, sinusoidal, and shock responses) that occur during launch.[17] Likewise, CubeSat structural analysis allows for defining the material from which it will be manufactured, based on the results obtained from the simulation of dynamic stresses.[18]

This study presents a dynamic computational structural analysis of the CIIASaT CubeSat structure designed by the Center for Research and Innovation in Aeronautical Engineering.

2 Methodology

The CIIASaT design considered a new experimental material to protect the payload and subsystems within the spacecraft against space radiation at Low Earth Orbit. This new material must be applied on large surface areas, imposing initial restrictions on the structure.

For the design and analysis of the CIIASaT structure, a method was created that satisfied the design and analysis specifications for space structures; this method was divided into four phases.[19,20]

- *Phase I (planning and clarification).* This phase determined the properties, restrictions, and characteristics of a CubeSat (Fig. 1a).
- *Phase II (conceptual analysis).* In this phase, the conceptual mathematical models and structural dynamics were evaluated. The structure was modeled in Autodesk Inventor software (Fig. 1b).

[15] Cıhan M., C̦ etın A., Kaya M. and Inalhan G., Design and analysis of an innovative modular cubesat structure for itu-psat ii, in Proceedings of 5th International Conference on Recent Advances in Space Technologies—RAST2011. pp. 494–499, https://doi.org/10.1109/RAST.2011.5966885, 2011.

[16] Steven H. and Huzain M., Requirements and design structure for surya satellite-1, IOP Conference Series: Earth and Environmental Science 149: 012063. https://doi.org/10.1088/1755-1315/149/1/012063, 2018.

[17] Alhammadi A., Al-Shaibah M., Vu AAT, Tsoupos A., Jarrar F. and Marpu P., Quasi-static and dynamic response of a 1u nano-satellite during launching, in 8th European Conference for Aeronautics and Space Sciences (EUCASS), https://doi.org/10.13009/EUCASS2019-398, 2019.

[18] Sekerere K. and Mushiri T., Finite element analysis of a cubesat, in International Symposium on Industrial Engineering and Operations Management, Bristol, UK. URL http://www.ieomsociety.org/ieomuk/papers/32.pdf, 2017, [6 July 2022].

[19] Bermudez-Reyes B. and Arroyave J. H., (IMPI MX/a/2016/005549, patent under revision, Méx-

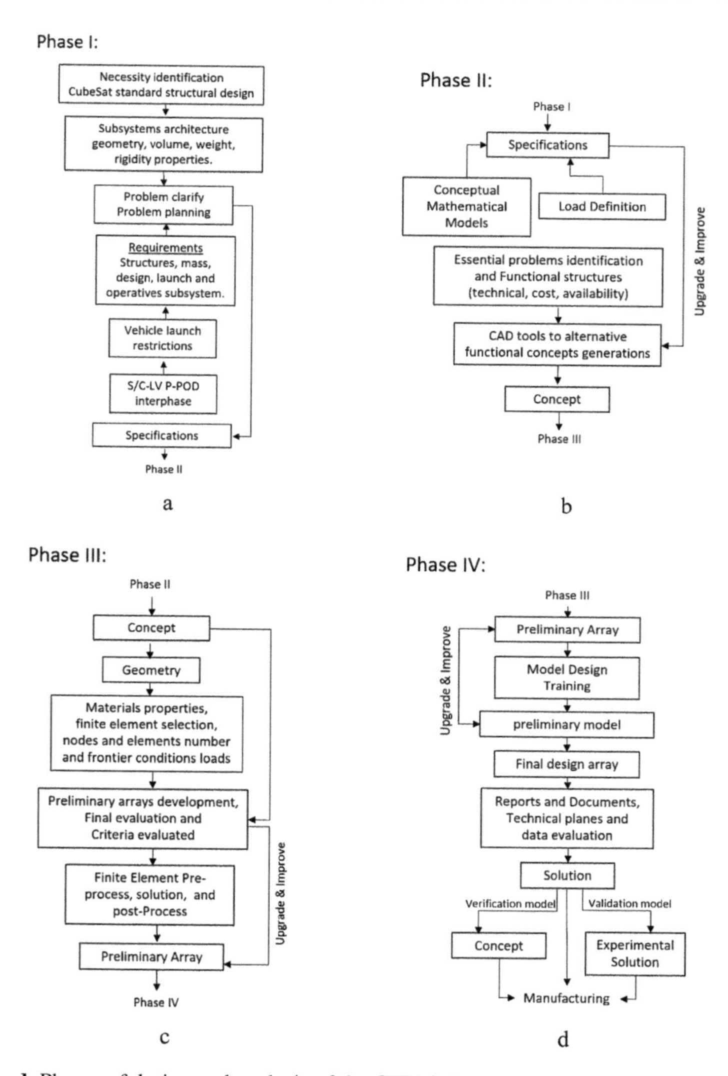

Fig. 1 **a–d**: Phases of design and analysis of the CIIASaT

ico) Nanosatélite tipo "cubesat", 6 July 2022, https://vidoc.impi.gob.mx/visor?usr=SIGA&texp=SI&tdoc=E&id=MX/a/2016/005549, 2017.

[20] Herrera-Arroyave J., Bermúdez-Reyes B., Ferrer-Pérez J. and Colín A., Cubesat system structural design, in 67th International Astronautical Congress (IAC), Guadalajara, México: International Astronautical Federation (IAF), 2016.

Table 1 CIIIASaT structural characteristics

Structural characteristics			
Component	Quantity	Material	Mass (kg)
Body	1	6061-T6	0,13542
Bottom plate	1	6061-T6	0,05506
Top plate	1	6061-T6	0,05467
Integrator shaft	4	Ti-6Al-4V	0,01465
Seeger	12	Ti-6Al-4V	0,00027
Total	**19**		**0,256885**

- *Phase III (preliminary analysis).* This phase comprised simulating and analyzing the CubeSat structure in ANSYS Multiphysics and R15.0 using the data derived in Phases I and II (Fig. 1c).
- *Phase IV (detailed analysis).* In this phase, all the planes and assembly procedures were generated alongside the verification, validation, and manufacturing of the CubeSat (Fig. 1d).

3 Results and Discussion

This section provides the primary results derived using the methodology described above.

Phase I

The CIIIASaT's structural design met the required size and shape requirements, the exterior dimensions, and the designated design reference frame and orientation (see Table 1).

According to the CalPoly standards for designing CubeSats, the following requirements must be considered:

- *Structure.* The structure must protect the payload and be easy to assemble. This implies the selection of specific materials.[21]
- *Mass.* The structural mass must not exceed 0,3 kg.

[21] Al-Maliky F. and AlBermani M., Structural analysis of kufasat using ansys program, Artificial Satellites 53(1): 29–35, 2018.

- *Design.* The structure must provide easy access to the payload and must consider the (P-POD[22]) structure. For the CubeSat developed in the current study, Ariane V5 was selected as the launch platform.[23,24]
- *Launch.* The structure must support static acceleration. It must not experience vibration based on power spectral density (PSD). The random vibration spectrum, sinusoidal spectrum, shock reply spectrum, and acoustic vibration spectrum must be performed according to the selected launch vehicle.[25,26,27]
- *Operational.* According to CubeSat standards, all the structural parts must keep joint during the launch discharge and throughout the procedure.
- *Geometry.* The structure must comply with size and geometric requirements, (cubic geometry, $10 \times 10 \times 10$ cm). This is equivalent to 1 U ("unit").[28]
- According to Table 1, CIIIASaT weighs 0,03993 kg less than the CubeSat standard. This allows for adding an additional electronic array to the payload or a passive thermal insulator system. With these additions, it is possible to calculate the natural frequencies and altitudes system, architecture determination, and placement into launch vehicle.[29]

Phase II

The coordinating global reference system used to perform static and dynamic analyses of the CIIIASaT CubeSat structure is shown in Fig. 2.

It is observed that (0, 0, 0) coordinated is localized in the center of the low plate surface, where the +X-axis is on the front surface structure. The +Y-axis was located at the right lateral face and the +Z-axis at the top face. It was parallel to the launch vehicle's longitudinal axis or positioned in the same launch direction. The CIIIASaT primary structure was designed with an external body and integrated axes (Fig. 3) comprising one solid-walled body, two cover plates, four integrated shafts, and eight Seeger rings as fixed elements. This solid walled body presented

[22] P-POD stand for Poly-Picosatellite Orbital Deployer.

[23] CalPoly, Picosatellite Orbital Deployer User Guide, Technical report California Polytechnic State University, 2014.

[24] NASA (2016), Standard Materials and Processes Requirements for Spacecraft NASA-STD-6016A, Standard, National Aeronautics and Space Administration.

[25] NASA (2020), NASA State of the Art Small Spacecraft Technology, 6 July 2022, https://www.nasa.gov/smallsat-institute/sst-soa.

[26] DOD, Product Verification Requirements for Launch, Upper Stage, and Space Vehicle MIL-STD-1540D, Standard, Department of Defense, 1999.

[27] NASA (2001), NASA Technical Handbook: Dynamic Environmental Criteria, NASA-HDBK-700, Handbook, National Aeronautics and Space Administration.

[28] NASA (2013), NASA General Environmental Verification Standard (GEVS): For GSFC Flight Programs and Projects NASAGSFC-STD-7000A, Standard, National Aeronautics and Space Administration.

[29] NASA (2014), NASA Launch Service Program, Program Level Dispenser and CubeSat Requirements Document, NASA-LSPREQ-317.01-Rev-B. Handbook, National Aeronautics and Space Administration.

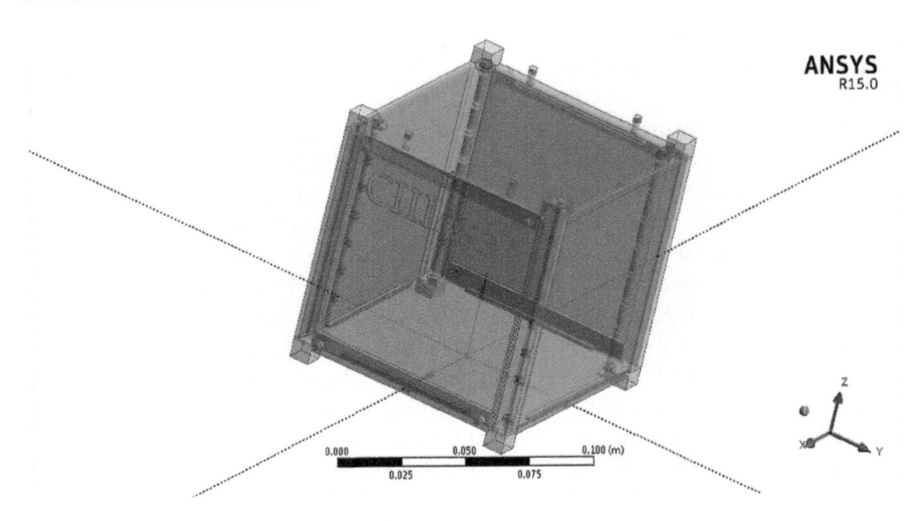

Fig. 2 CIIIASaT coordinated global reference system

sufficient space for coverage by this new material to protect the inner components against space radiation as required. Similar-shaped structures and different manufacturing techniques are being proposed for other missions (see Footnote 8). The structure in Fig. 2 has the advantage of being simple to manufacture and assemble.

Based on the above Phase I, the dynamic loads were established for the Ariane V5 rocket. The four loads were studied as follows: (1) quasi-static acceleration

Fig. 3 CIIIASaT primary structure parts

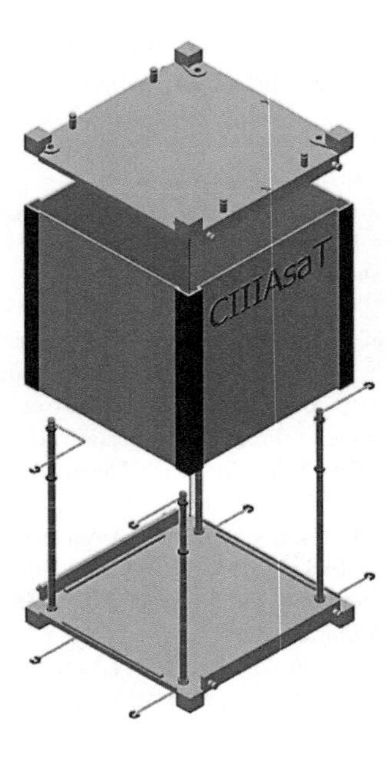

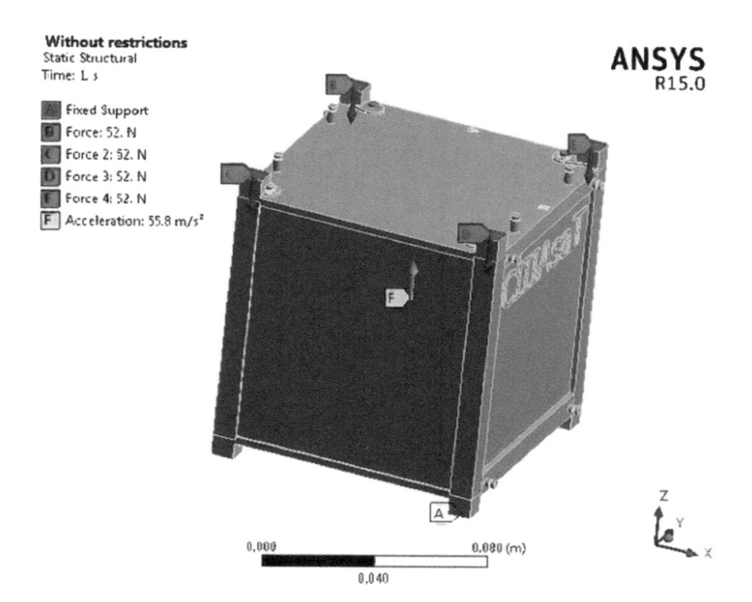

Fig. 4 Static forces distribution on CIIASaT

loads (static analysis), (2) random mechanics vibration loads (random vibration analysis), and (3) shock mechanics loads (response spectrum analysis). For each of these, a (4) modal analysis was conducted, except for the quasi-static loads. It is noted that structural joints were created using glide and Seeger rings. In this manner, screws were eliminated to avoid the disassembling of the nanosatellites caused by vibrations during launch (Footnotes 24 and 29).

Phase III

In this phase, the structural behavior at different spatial–mechanics loads and according to the launch vehicle was studied.[30]

Static Analysis: The loads that contributed to the weight of the other two satellites into the P-POD and the P-POD expulsion spring also had to be considered. Accordingly, the total static force that the CIIASaT could support was 222,93 N (Fig. 4).

According to Fig. 1c, the static force is distributed onto the four-top vertex; this implicated that each vertex supported 52,00 N. The low vertex was restricted in all degrees of freedom, which indicated that it was statically supported; this defines a frontier condition. To apply an inertial load, e.g., static launch acceleration, it is necessary to simulate the force that exceeds the CIIASaT's structural mass (14,52 N). The results of the static global analysis are summarized in Table 2. This

[30] ArianeSpace, Ariane 5 Users Manual Issue-5, 2011.

Table 2 CIIIASaT static global analysis results of Seeger ring 2

Structural characteristics of Seeger ring 2			
		Max	Min
Total deformation (mm)	δ_{total}	0,00358	0,00000
Directional deformation (mm)	δ_x	0,00265	−0,00274
	δ_y	0,00064	−0,00041
	δ_z	0,00000	−0,00358
Normal stress (MPa)	S_{xx}	3,39850	−4,45970
	S_{yy}	3,24240	−3,78240
	S_{zz}	2,73750	−6,99770
Shear maximum stress (MPa)	τ_{max}	3,82910	0,00000
Stress intensity (MPa)	l	7,65820	0,00000
Von Mises–Hencky equivalent stress (MPa)	σ_e	6,64180	0,00000
Principal stress (MPa)	σ_A	4,99580	−2,04240
	σ_B	2,22060	−4,15100
	σ_C	1,28210	−7,62360
Shear orthogonal stress (MPa)	S_{sxy}	2,66500	−2,63660
	S_{syz}	1,65610	−1,56130
	S_{sxz}	1,76170	−1,61140
Reaction force (N)	222,51		
Security factor	15		
Marginal security factor	14		

table includes the maximum and minimum values of the most important structural parameters.

Table 2 also shows the values of Seeger ring 2 (Fig. 5). Following the von Mises–Hencky failure theory, the fluency yield of a Ti6Al4V alloy allowed for obtaining a security factor of 15. Consequently, the CIIIASaT structure was able to endure the induced statistical loads on the launch vehicle. Conversely, the maximum deformation was 0,00357 mm on the low plate of the CIIIASaT.

Previous results were verified and compared with the ANSYS results; these were congruent and it was found that the loads acted correctly.

Modal Analysis: The first six frequencies obtained for the CIIIASaT structure are compared with the values obtained by King[31] in Table 3.

When comparing both frequencies, a maximum difference of 10% was detected. These differences resulted from the CubeSat's geometry and the material used to create it (both structures were designed based on CubeSat standards). Accordingly, it was possible to eliminate six degrees of freedom and the other part of the structure was free. Based on these conditions, 26 primary modal data results were

[31] King D., Nanosatellite Structure Design, Master's Thesis, York University, Canada, 2010.

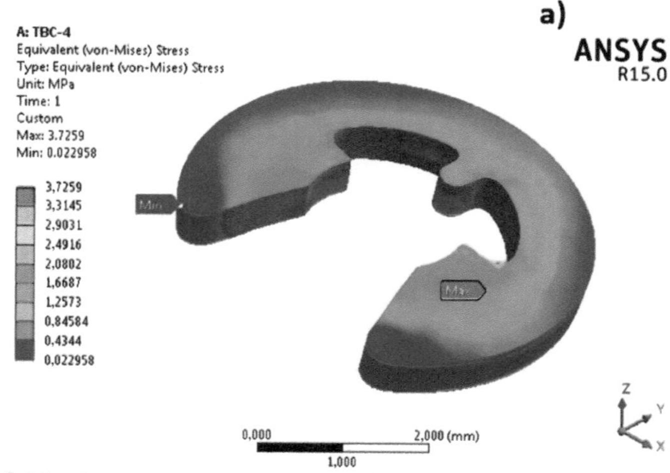

a) Maximum stress on seeger 2.

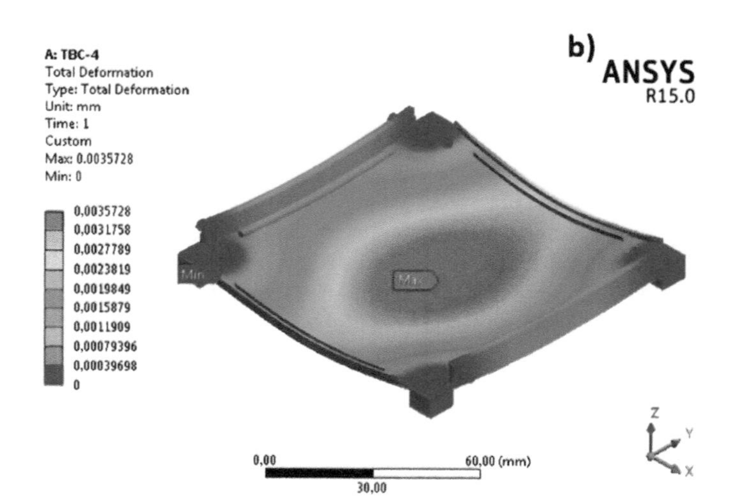

b) Maximum deformation on the low plate.

Fig. 5 Maximum stress on Seeger 2 and maximum low plate deformation

Table 3 Structural mode comparison between CIIASaT and King

Mode	CIIIAsaT (Hz)	King (see Footnote 31) (Hz)	%E
1	979,1	898,7	8,94
2	1.006,6	945,3	6,48
3	1.042,9	981,6	2,37
4	1.103,0	1.013,0	8,88
5	1.190,5	1.077,0	10,53
6	1.204,39	1.216,0	0,91
Total %Error	**19,0**		**6,35**

obtained for the X-axis. Mode 23 had a modal participation factor with greater amplitude (2.527 Hz) while Mode 92 (7.545 Hz) had a modal participation factor with a lower amplitude.

Modal shapes on the Y- and Z-axes were similarly determined. A total of 21 modal shapes were obtained for the Y-axis; Mode 25 had the highest dynamic modal participation factor (2.674 Hz) and Mode 159 had the lowest (12.243 Hz). Furthermore, dynamic modal shapes were determined for the Z-axis (Fig. 6), where Mode 80 represented the highest dynamic participation factor (6.489 Hz) and Mode 3 represented the lowest dynamic participation factor (1.043 Hz). Natural frequencies were classified from highest to lowest for all modal shapes in each excitation direction (X, Y, and Z).

Harmonic Response Analysis: The harmonic response analysis is a sinusoidal excitation transmitted by the launch vehicle to the CubeSat base. The sinusoidal spectrum was applied on the base of the structure, directly on the fixed support elements. By doing so, it was possible to extend the bandwidth up to 2.000 Hz because the first natural frequency of the structure was 979,144 Hz. Accordingly, the dynamic structural properties were obtained using this load. This did not affect the primary structure because the acceleration applied was very low compared with the quasistatic acceleration and response spectrum loads.

Random Vibration Analysis: The acoustic loads from the launch vehicle were transformed into random mechanical vibrations that affected the CubeSat base. The Ariane V5 operation manual does not specify an acoustic vibration; therefore, it was necessary to calculate the PSD excitation corresponding to a frequency ranging from 30 to 2.000 Hz to obtain the complete acceleration spectrum as follows:

$$w(f_y) = w(f_{\text{ref}})\left(\frac{f_y}{f_{\text{ref}}}\right)^{\frac{r}{3}}, \quad (1)$$

where

$w(f_y) = \text{PSD},$

Fig. 6 Modal shapes in Z-axis

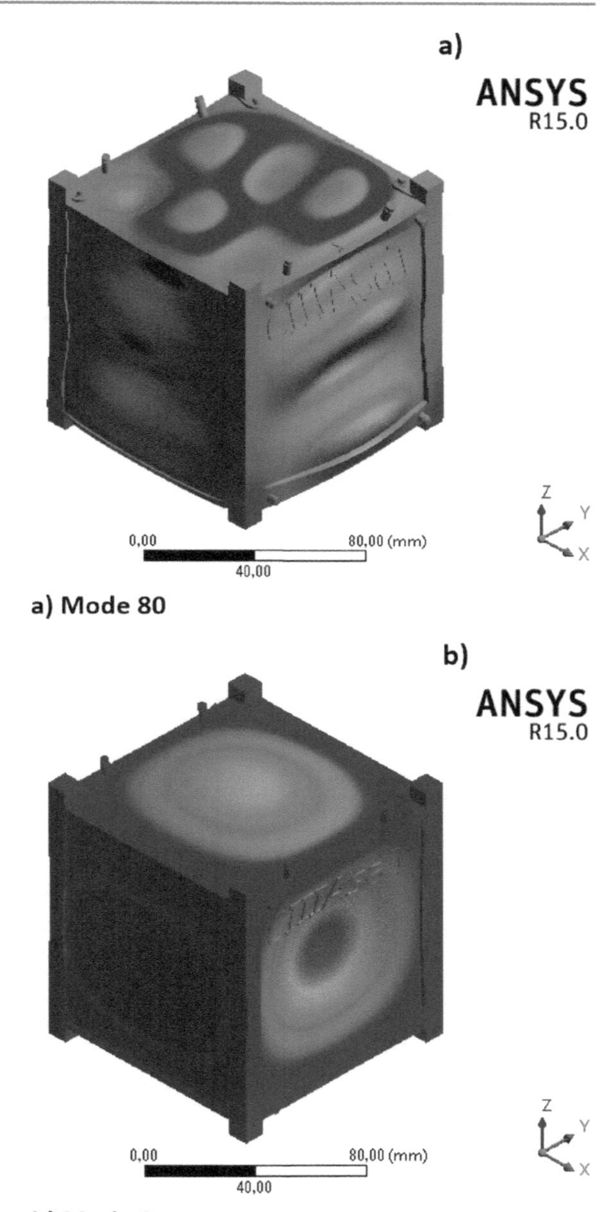

$w(f_{\text{ref}}) = 0{,}09\text{g}^2/\text{Hz}$ (acoustic noise spectrum density function: 150–700 Hz),

f_y: frequency on the y-axis,

f_{ref}: reference frequency,

r: octave bandwidth ($-3\,\text{dB/oct}$ for 2.000 Hz and $+\,6\,\text{dB/oct}$ for 30 Hz).

Substituting all the values in Eq. (1) the following was obtained:

$$w(f = 30) = 0{,}0036\frac{\text{g}^2}{\text{Hz}},$$

$$w(f = 2.000) = 0{,}0315\frac{\text{g}^2}{\text{Hz}}.$$

These results were used to determine the PSD (Fig. 7). Thus, to calculate the mean root response value of PSD, ($\overline{R} = x_{\text{rms}}$) was divided into three areas (A_1, A_2, and A_3). According to these areas of PSD graph were calculated as follows:

$$\text{A}_1 = \frac{w_2 f_2}{n+1}\left[1 - \left(\frac{f_1}{f_2}\right)^{n+1}\right], \tag{2}$$

$$\text{A}_2 = w_2(f_3 - f_2), \tag{3}$$

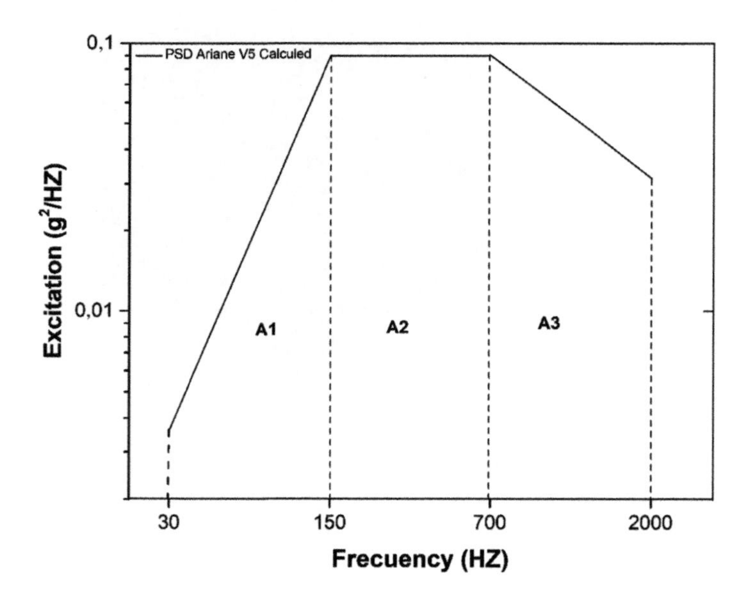

Fig. 7 PSD for Ariane V5 calculated

$$A_3 = \begin{cases} \frac{w_3 f_3}{m+1}\left[\left(\frac{f_4}{f_3}\right)^{m+1} - 1\right], m < 0, m \neq -1, \\ w_3 f_3 ln\left(\frac{f_4}{f_3}\right), m = -1. \end{cases} \tag{4}$$

where

$$\begin{aligned} n &= \frac{r_{1\to2}}{3} = \frac{+6\,\text{dB/oct}}{3}, \\ m &= \frac{r_{3\to4}}{3} = \frac{-3\,\text{dB/oct}}{3}, \\ f_1 &= 30\,\text{Hz}, \\ f_1 &= 150\,\text{Hz}, \\ f_1 &= 700\,\text{Hz}, \\ f_1 &= 2.000\,\text{Hz}. \end{aligned}$$

Substituting values in Eqs. (2)–(4) yielded $A_1 = 4{,}464\ g^2$, $A_2 = 49\ g^2$, $A_3 = 66{,}138\ g^2$. Then, $\overline{R} = x_{rms} = \sqrt{A_1 + A_2 + A_3} = 10{,}92\,g$. This value is the same as the RMS for Ariane V5 (11 G). Therefore, the results obtained for the dynamic structural response of CIIIASaT due to random vibration excitation were defined as located on the Z-axis. According to the response spectra, each of the three events on the CIIIASaT structure was obtained the load due to separation. This behavior affected more than the excitation load adaptor separation, where the maximum total deformation was on the low plate of the structure (0,5399 mm). This deformation was not sufficient for the aluminum plate to present a plastic deformation (Fig. 8).

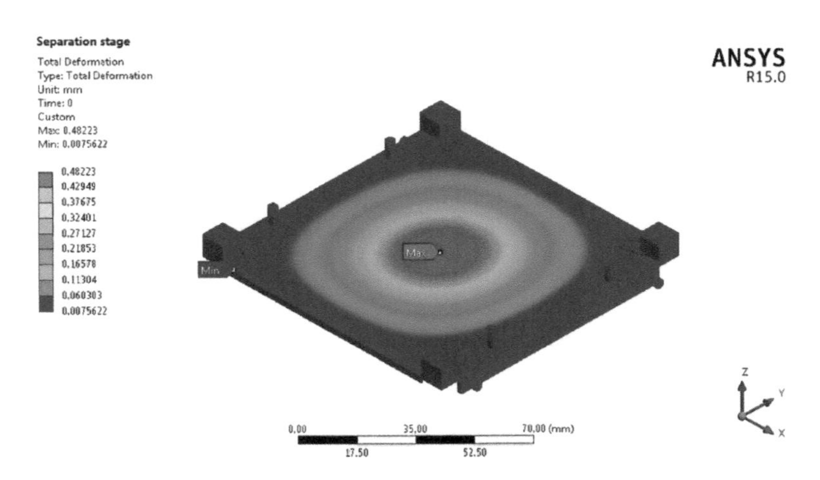

Fig. 8 Total deformation of response spectrum

Similarly, the von Mises–Hencky equivalent stress was obtained from the response spectrum for the first stage and fairing separation, i.e., maximum equivalent stress of 1.113 MPa. This stress was localized on Seeger ring 2. Compared with the Ti4Al6V last stress (1.949 MPa), the Seeger ring 2 resisted the mechanical–spatial conditions.

Response Spectra Analysis for each one of the events on the CIIIASat structure, the load due to separation according to the response spectra was obtained. This behavior affected more than the excitation load adaptor separation, where the maximum total deformation was on the low plate of the structure (0,5399 mm). Accordingly, this ring resisted the mechanical-spatial conditions. These values were compared with those obtained by Orr et al.[32] and Pierlot[33] for their respective nanosatellites and were the same.

The maximum longitudinal acceleration was produced for Ariane V5 during the impulse of the cryogenic stage and did not exceed 4,55 G. The highest static lateral acceleration was 0,25 G. This meant that the maximum acceleration occurred during Phase 1, just after the principal cryogenic engine had been initiated (see Footnote 30). Therefore, the total force applied longitudinally to the structural body was the most critical static load that the structure had to endure during the launch (see Footnote 25). According to von Mises–Hencky' equivalent stress value, this did not exceed 6,6769 MPa. This value represented <5% of the Al-6061T6 and Ti6Al4V elongations. These materials are thus recommended for use in aerospace manufacturing (see Footnotes 24 and 29).

In the design stage, the natural frequencies and modal forms of the CIIIASaT were determined for bandwidth ranging from 0 to 1.500 Hz. In this range, 200 modes were obtained. These mode numbers were sufficient for performing complete dynamic structural analysis and validation tests using the ANSYS software. The harmonic response did not exceed the average acceleration value of 1 G in the Z-direction for the 20–100-Hz bandwidth with a security factor of 1,25.[34,35] When these results were compared against those of King (see Footnote 31), a large similarity in modal frequencies was observed for both structures. A minimal difference (6,35%) in the total error percentage was obtained between the two structures. Nevertheless, they were characterized because both met the standard size of a CubeSat.

The current modal study illustrated an analytical system that included the geometric and material properties of each component of a CubeSat and the relationship

[32] Orr N., Ayer J., Larouche B. and Zee R., Precision formation flight: The can x-4 and can x-5 dual nanosatellite mission, in 21st Annual AIAA/USU Conference on Small Satellite. AIAA, 2017.

[33] Pierlot G., Oufti-1: Flight System Configuration and Structural Analysis. Master's Thesis, Aerospace, and Mechanical Engineering Department, University of Liége, Belgium, 2009.

[34] Abdelal G., Abuelfoutouh N. and Gad A., Finite element analysis for satellite structures: applications to their design, manufacture and testing, Springer-Verlag London. ISBN 978-1-4471-4636-0, https://doi.org/10.1007/978-1-4471-4637-7 2013.

[35] Wijker J., Spacecraft structures, Springer-Verlag Berlin Heidelberg, ISBN 978-3-540-75552-4, https://doi.org/10.1007/978-3-540-75553-1, 2008.

between them.[36] It is noted that to achieve these results, the effective mass as a participation factor related to friction and assembly was considered (see Footnote 31). In this way, the different modes for different directions were obtained within the same bandwidth (0–1.500 Hz). The CIIASaT's structural analysis used the square root of the sum of the square's method. To perform this technique, three response spectra points were used as follows. (1) The excitation was applied as a response spectrum; for each spectrum value, there is a corresponding frequency. (2) The excitation was applied for fixed supports. (3) The excitation response was calculated from 200 modes that were derived from the modal analysis using ANSYS.[37]

The CIIASaT structure will be subject to different shock loads during several events as follows: (1) upstage separation of the cryogenic main stage; (2) fairing separation; (3) main payload separation; (4) adaptor separation. The shocks generated by stages (1) and (2) are propagated from the origin until the payload base through the vehicle structure. Shocks produced by payload separation will be generated directly on the main load base and their levels will depend on adaptor type (see Footnote 30).

Phase IV

For computational verification and validation for the CIIASaT, a simulation with a static analysis focus was developed to find the total maximum deformation of the structure. Thus, twelve simulations were conducted, each verifying the size of the elements. In all simulations, the maximum deformation was detected at the low plate of the structure. A graph was derived based on deformations and node data for all the simulations (Fig. 9). Moreover, it was observed that meshes presented independence in terms of the results obtained. The total deformation converged at 3,31 μm.

Case study validation, based on comparing the data obtained from a mathematical model with that obtained by a structural computational model was conducted.[38,39] Finally, the CIIASaT mathematical model was established in MATLAB and was compared with the ANSYS CIIASaT computational model (see Table 4).

According to Table 4, it is possible to compare the modal frequencies obtained in the computational and mathematical models. The error derived from this

[36] Hatch M. Vibration simulation using MATLAB and ANSYS, Chapman and Hall/CRC, 2000.

[37] ANSYS W., 15.0 User's Guide, Canonsburg, PA, USA, 2014.

[38] Brieß K., Spacecraft Design Process, Chapter 8, Chichester, United Kingdom: John Wiley and Sons, pp. 647–737, https://doi.org/10.1002/9780470742433.ch8, 2009.

[39] AIAA, AIAA Guide for the Verification and Validation of Computational Fluid Dynamics Simulations, AIAA G.: American Institute of Aeronautics and Astronautics, ISBN 9781563472855, https://doi.org/10.2514/4.472855.001, 1998.

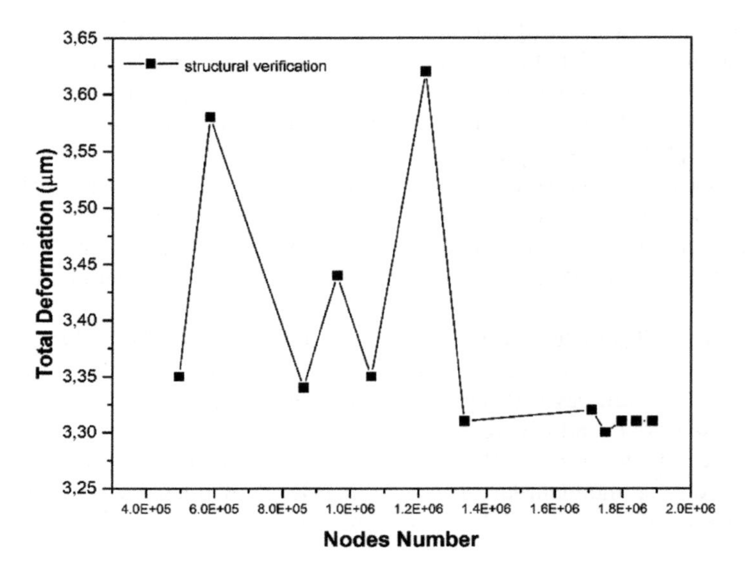

Fig. 9 Structural verification chart of CIIIASaT Low plate maximum deformation

Table 4 Validation data results

	Modal frequency obtained (Hz)			% Error	
Mode	M.[a]	MAT.[b]	AN.[c]	M. versus MAT.	M. versus AN.
1	15,42	15,60	15,63	1,15	0,19
2	96,65	97,74	98,13	1,11	0,40
3	270,60	273,80	276,30	1,16	0,90
4	530,30	537,08	543,10	1,25	1,11
Average				1,17	0,65
Std. dev.				0,05	0,37

[a]Mathematical model
[b]MATLAB
[c]ANSYS

comparison was <1,17% with a standard deviation of 37%. Therefore, the computational analysis of the CIIIASaT structure can be considered convergent, correct, and independent of the mesh (see Footnotes 39 and [40]).

[40] Oberkampf W. and Roy C., Verification and validation in Scientific Computing, Cambridge University Press, https://doi.org/10.1017/CBO9780511760396, 2010.

4 Conclusion

In the structural and dynamic analysis of the CIIASaT structure, the influence of modes was considered to find the effective mass, which was used to classify the importance relative to the mode. The fundamental frequency of the CIIASaT structure was determined as 979,14 Hz, which was not in the bandwidth of the harmonic vibration spectrum (2–100 Hz). This indicated that the structure will not resonate with the launch vehicle during the flight into orbit. The design met the mechanical-spatial requirements according to the load levels for acceptance tests. The CIIASat structure does not have threaded fasteners. Therefore, no dissembling of the structure is anticipated when performing the different tests in mechanical–spatial conditions for the selected launch pad.

***

Bárbara Bermúdez-Reyes She is a Graduate Level Lecturer at the School of Engineering of the National University Autonomous of Mexico. Previously she was a Research Professor at the Faculty of Engineering Science and Technology of the Autonomous University of Baja California. She was General Secretary of the University Space Engineering Consortium-Mexico Chapter for the period 2014–2019. She is a member of the Space Science and Technology Network. She has collaborated on projects with Mexican Air Force and has evaluated projects of the Mexican Naval Research Institute. She is part of the organizing committee of the 6th National Cansat Contest 2021. Her research lines are the design and processing of aerospace ceramic-ceramic, metal-ceramic composites materials and satellite structures.

Jorge Enrique Herrera Arroyave . He was born in Pereira, Colombia. He is currently a Research Professor at Pereira Technological University since 2016 and has been a Professor of Engineering at the Faculty of Basic Sciences and Engineering of the Catholic University of Pereira, Risaralda, Colombia. His research has been related to machine design, structural dynamics, and renewable energy. He is Mechanic Engineer and he has a postgraduate in Aeronautics Engineering Science.

Patricia Zambrano Robledo She is a Senior Research Professor in the Research and Innovation Center in Aeronautical Engineering of the Autonomous University of Nuevo Leon. She is a faculty member of the Materials Science Doctorate program and Aeronautical Doctorate Program. Her research areas of interest include the processing and characterization of metal- and ceramic-matrix composites (MMCs and CMCs), as well as superalloys and Aadditive mManufacturing. She has published around 50 papers in international journals and about 100 articles at international conferences. She is the author or co-author of about 6 book chapters. Since January 2016 she is a Research Director at UANL (Autonomous University of Nuevo León), Mexico.

Rafael Vargas-Bernal He received his D.Sc. in Electronic Engineering from the INAOE, Tonantzintla, Puebla, Mexico, in 2000. He is an associate professor in the Materials Engineering Department from Instituto Tecnologico Superior de Irapuato, in Irapuato, Guanajuato, Mexico. He is a researcher of the National System of Researchers from Mexico with level I. Also, he belongs to the research group called "Advanced Materials Applied to Engineering". He has been a reviewer in journals for RSC, Elsevier, and IEEE. He has published 13 articles in indexed journals and 32 chapters in books. His areas of interest are nanomaterials, aerospace materials, composites, and gas sensors.

Jorge Alfredo Ferrer Pérez He is an associate professor at the National University Autonomous of Mexico-School of Engineering. He received his Ph.D. in Aerospace and Mechanical Engineering from the University of Notre Dame, South Bend in the United States. He is part of the Aerospace Engineering Department and is responsible for the Space Propulsion and Thermo-vacuum lab. This facility belongs to the Space and Automotive Engineering National Laboratory located at Juriquilla. Querétaro. His current research areas are nano-heat transfer in solid-state devices, thermal control, space propulsion, small satellites, and the development of space technology.

Lessons Learned on the Thermal Analysis of a Cubesat Using the Finite Element Method

Dafne Gaviria-Arcila, Jorge A. Ferrer-Pérez, Carlos Romo-Fuentes, Rafael G. Chávez-Moreno, Jose Alberto Ramírez-Aguilar, and Marcelo López-Parra

Abstract

CubeSats have been gaining importance over the last years because of their short time of development, small budgets, and accessible components to be acquired or manufactured. In Latin America, this kind of spacecraft has been popular, and more projects are known to be in progress. CubeSats are formed by several subsystems which are conceived depending on the mission objective. The Thermal Control System (TCS) is responsible to maintain the operative temperature range of all the components within the spacecraft. This subsystem interacts with the others spacecraft's subsystems and is usually underestimated. To design the TCS, an iterative process is followed to predict the CubeSat temperature distribution as a function of time along its orbit and space within the spacecraft components. By using numerical approaches such as the Finite Element Method, Finite Volume Method, and Finite Difference Methods it is possible to solve the differential equations to determine the temperature field. Although several analyses have been reported in the literature, implementation details are

D. Gaviria-Arcila · J. A. Ferrer-Pérez (✉) · C. Romo-Fuentes · R. G. Chávez-Moreno · J. A. Ramírez-Aguilar · M. López-Parra
Advanced Technology Unit, School of Engineering, UNAM, Juriquilla, Querétaro, Mexico
e-mail: ferrerp@unam.mx

D. Gaviria-Arcila
e-mail: dafne.gaviria@comunidad.unam.mx

C. Romo-Fuentes
e-mail: carlosrf@unam.mx

R. G. Chávez-Moreno
e-mail: rchavez@comunidad.unam.mx

J. A. Ramírez-Aguilar
e-mail: albert09@unam.mx

M. López-Parra
e-mail: lopezp@unam.mx

A. Froehlich (ed.), *Space Fostering Latin American Societies*, Southern Space Studies,
https://doi.org/10.1007/978-3-031-20675-7_7

not presented. This manuscript has the objective to show the lessons learned in obtaining the temperature distribution for the thermal analysis of a CubeSat using a practical case called K'OTO. During the thermal analysis of K'OTO, there were identified different challenges to perform the numerical modeling and to predict the temperature of the CubeSat. Some of the challenges presented were the definition of thermal and optical material properties, the simplification of geometry, the definition of contacts among components, the preparation of thermal loads, the change in temperature over time, and the effects of the internal component distribution on the variation of temperature. The main purpose of this work is to provide the "know-how" of CubeSat thermal numerical simulation using Finite Element Analysis (FEA) to interested readers in Mexico, Latin America, and other parts of the world showing critical hidden details that usually are not reported or presented.

1 Introduction

The main function of the Thermal Control System (TCS) is to manage the temperature of a satellite and its internal components. The distribution of the temperatures of and inside a satellite depends on the heat transferred by radiation from the space environment to the satellite and vice versa and the heat transferred by conduction between the satellite structure and the internal components. The external heat loads from the space environment are from the solar flux, the infrared emitted by Earth (IR Earth Flux), and the Albedo flux as it is presented in Fig. 1.[1]

The satellite exposition to the solar flux is affected by the season of the year, the distance between the satellite and the Sun, and the inclination of the satellite. The Albedo Flux is the reflection of the solar radiation on other planets. Thus, the Albedo flux depends on the angle of incidence of solar radiation over the satellite. The IR Earth Flux is a function of the position and orientation of the Earth and the satellite orbit height.[2]

The internal heat loads of the satellite rely on the heat energy dissipated from the internal components such as circuits, processors, or Commercial off-the-shelf (COTS) components. The internal heat dissipation is basically due to the energy consumption and the resistance between the joints of these components.

[1] Ferrer-Pérez, J.A., Gaviria-Arcila, D., Romo-Fuentes, C., Chávez-Moreno, R.G., Ramírez-Aguilar, J.A. and López-Parra, M. The Development of CubeSats in Latin America and Their Challenges on the Design of Thermal Control Systems, in: Froehlich, A. (eds) Space Fostering Latin American Societies. Southern Space Studies. Springer, Cham. https://doi.org/10.1007/978-3-030-97959-1_3, 2022.

[2] Ovchinnikov, M., Mckenna-lawlor, S., Psychology, S., Kanas, N., Manuy, D., Larson, W. J., and Wertz, J. R., Space mission analysis and design, in Choice Reviews Online, Vol. 29, Issue 09, https://doi.org/10.5860/choice.29-5149, 1992.

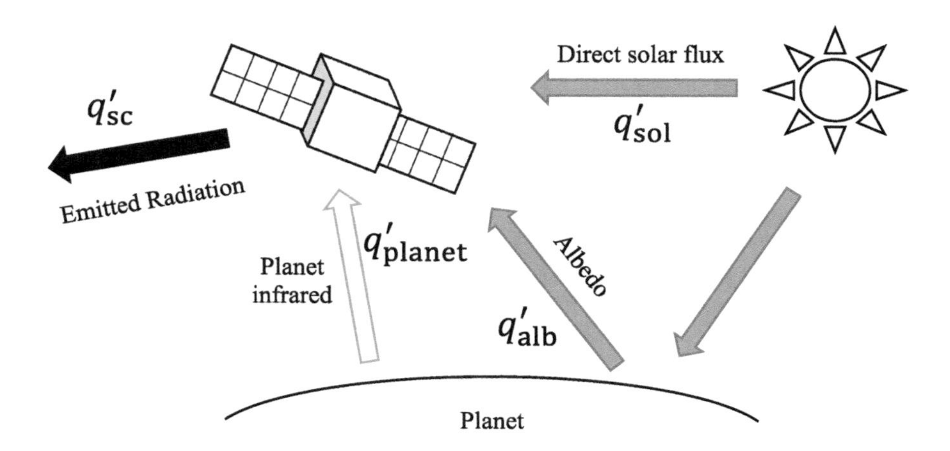

Fig. 1 Heat transfer between the space environment to the satellite and vice versa

The knowledge of the external and internal heat loads involved in the satellite is part of the analysis of the TCS to propose a reliable design. This is because the TCS should be able of supporting these loads in all operational modes of the satellite and all the mission stages.[3]

The TCS is sensible to various parameters due to the different mission stages, namely, the main parameters to consider are as follows:

- the satellite size,
- mass,
- power consumption,
- Keplerian elements, and
- altitude control details.

The TCS design relies on the variation of these parameters and the operational modes of each subsystem. Once these parameters are established in depth, the responsible thermal control engineer can propose an adequate TCS.

The TCS can be passive or active. The active TCS is the one that consumes power to manage the temperature of the satellite. The passive TCS consists of insulating materials, coatings, or paints to isolate the satellite from the external environment. The selection of these materials is based on the optical properties and the relation between absorptivity and emissivity in such a way that they favor satellite thermal management.[4] Therefore, the active TCS is proposed once the

[3] Baturkin, V., Micro-satellites thermal control—concepts and components, Acta Astronautica, 56(1), pp. 161–170. https://doi.org/10.1016/j.actaastro.2004.09.003, 2005.

[4] Fortescue, P. W., and Stark, J. P. W., Spacecraft Systems Engineering, West Sussex: John Wiley & Sons Ltd, 2003.

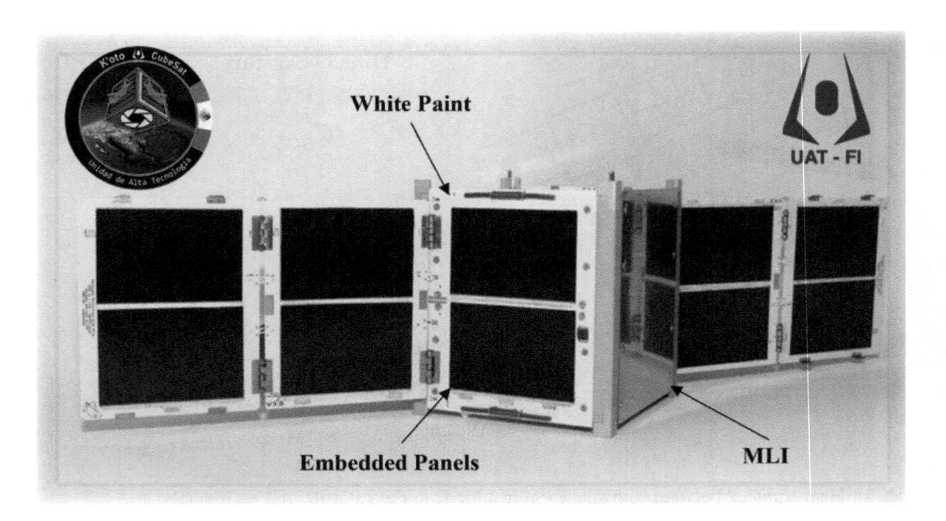

Fig. 2 Passive TCS was used in the design of the TCS of the K'OTO CubeSat

passive TCS is analyzed and verified i.e., in case the passive TCS itself is not enough to keep the satellite components within the temperature operational limits.

One example of the usage of the passive TCS is in the design of the TCS of the K'OTO CubeSat as shown in Fig. 2. The K'OTO TCS system use coatings such as white paint to cover the structure, Multi-Layer Insulation (MLI) on four faces of the satellite, and the use of embedded panels on the other two. The embedded panels act as a shell and insulate the satellite from the space environment.

K'OTO is a 1U CubeSat developed by the Advanced Technology Unit (UAT) from the UNAM and the Sustainable Development Secretariat (SEDESU) from the Government of the State of Querétaro, México. The main goal of the mission of K'OTO is to take imagery (photographs) of the Mexican territory and train future Aerospace Engineers in Mexico. This work presents the lessons learned during the development of the thermal analysis and design of the TCS of K'OTO.

This manuscript aims to present step by step the main challenges faced to carry out the thermal analysis of a CubeSat and to present lessons learned and recommendations to reduce the learning curve of the thermal analysis of future missions. The discussions of the lessons learned during the thermal analysis of K'OTO are presented in the following next sections.

2 Main Challenges of the TCS Analysis and Design

The analysis and design of the TCS face different challenges such as the insufficiency of available specific data as is the thermal performance of the COTS components.[5] Moreover, other peculiarities found during the thermal model preparation are:

- The definition of exact thermophysical properties.
- The definition of exact optical properties.
- The definition of realistic values of thermal contact in joints.
- The simplification of geometry for the TCS analysis.
- The preparation of thermal loads and choice of heat transfer mechanisms.
- The ability to trace the mechanical, structural, and internal component distribution according to the functionalities required in each stage and course of the project and the affectations on the thermal model.

3 Advantages of the Numerical Analysis

Over the years, technology has evolved and with it the ability to solve complex problems to represent physical phenomena i.e. computer-aided engineering (CAE). One application of the use of CAE is the thermal analysis of small satellites. The main advantages of CAE tools in the design of the TCS in a satellite are.

- to model the thermal balance in a satellite which includes the thermal loads coming from the space environment, the thermal energy dissipation of the internal components, and the radiation emitted by the satellite surfaces due to the optical properties of the materials,
- to predict the temperature distribution to iteratively adjust the design before making a physical prototype and provide visibility of details in places where it is not feasible or is difficult to access due to the miniaturization of the components and
- to reduce time and costs concerning the experiments due to the number of iterations that must be done to obtain a reliable design that guarantees the success of the satellite mission.

Figure 3 shows the summary of the advantages of CAE tools on the TCS design.

4 Numerical Methods to Analyze the TCS in a Satellite

The main goal to perform a thermal analysis in a satellite is to predict the distribution of temperature and thus identifying the components at risk while not meeting the thermal requirements. There are various numerical methods to predict the thermal performance of a satellite, namely, the Finite Element Method (FEM), the

[5] See footnote 3.

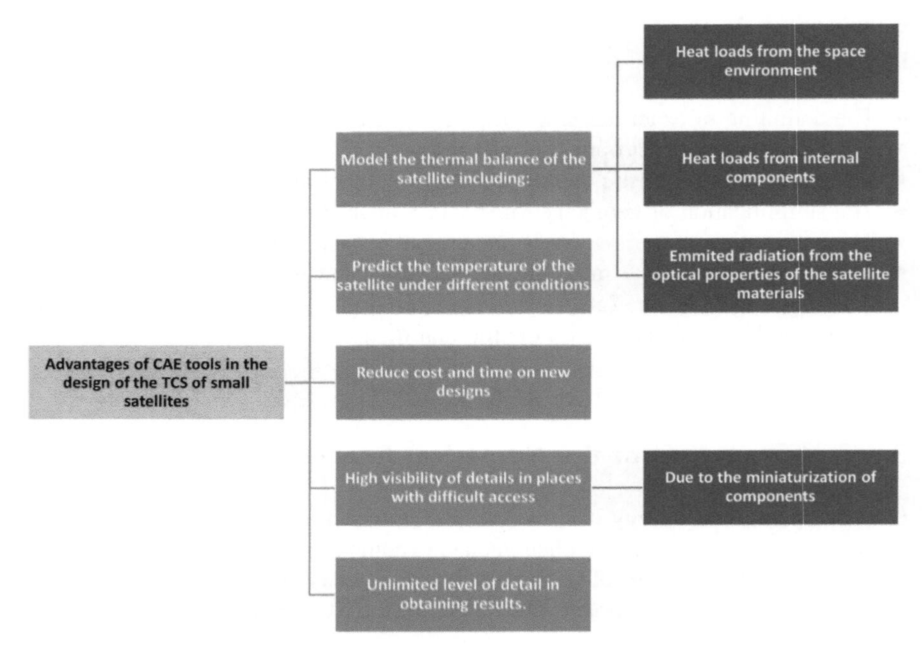

Fig. 3 Advantages of CAE tools on the TCS design

Finite Differences Method (FDM), and the Lumped Parameter Method (LPM). In general, these methods solve a partial differential equation that represents the heat transfer from the environment to the satellite and vice versa. This mathematical model is presented in Eq. 1 as follows,[6,7]

$$\underbrace{C\frac{dT}{dt}}_{\substack{\text{internal-energy}\\ \text{change rate}}} - \underbrace{q'_{sc}}_{\substack{\text{radiation from the}\\ \text{satellite to the}\\ \text{environment}}} - \underbrace{q'_{gen}}_{\substack{\text{heat rate generated by}\\ \text{the satellite}\\ \text{internal components}}} = \underbrace{q'_{sol} + q'_{alb} + q'_{planet}}_{\substack{\text{radiation from the}\\ \text{environment to the}\\ \text{satellite}}} \tag{1}$$

where the first term from the Left-Hand Side (LHS) is the internal-energy change rate. The second term and the third term from the LHS represent the radiation from the satellite to the space environment and the radiation caused by the dissipation of thermal energy from the internal components of the satellite, respectively. Both these last terms contribute to the thermal energy transferred from the satellite to

[6] Gilmore, D. G., and Donabedian, M., Spacecraft thermal control handbook (2nd ed.), Aerospace Press, 2002.
[7] Miao, J., Zhong, Q., Zhao, Q., and Zhao, X., Spacecraft Thermal Control Technologies, P. Ye (ed.), Springer, http://www.springer.com/series/16385, 2021.

the surrounding space environment. Note that q'_{sc} is function of $F_{sat \to p}$ that is the view factor between the satellite and the planet. The terms on the Right-Hand Side (RHS) of Eq. 1 are the thermal loads imposed on the satellite by the space environment, where q'_{sol} is the solar heat rate, q'_{alb} is the albedo heat rate, and q'_{planet} is the infrared radiation emitted by Earth.

The numerical solutions discretize the domain i.e., divide the domain into small elements where each element represent an equation and for each element, we know the variables of interest of immediate neighboring elements. The main goal of discretization is to reduce a partial differential equation problem to a system of algebraic equations. Once the element size is selected, a network is defined where algebraic equations will be solved.

Equation 1 can be written for each node as follows (Eq. 2)[8]:

$$C_i \frac{dT_i}{dt} = q'_{sol} + q'_{alb} + q'_{planet} + q'_{gen} + \sum_{j=1}^{n} K_{ij}(T_j - T_i) + \sum_{j=0}^{n} R_{ij}\left(T_j^4 - T_i^4\right) \tag{2}$$

where C_i is thermal capacitance, and K_{ij} and R_{ij} are the conductive and radiative links or couplings between nodes respectively. The term $\sum_{j=1}^{n} K_{ij}(T_j - T_i)$ represents the heat conduction contribution, and $\sum_{j=0}^{n} R_{ij}\left(T_j^4 - T_i^4\right)$ the heat radiation.

In general terms, the FEM solves the general Eq. 2 reducing from a differential equation to a system of algebraic equations This reduction is made using a piecewise polynomial approximation of the variable of interest (in this case the temperature of the satellite) in each element.[9] The FEM seeks for explicit solutions of the temperature which should satisfy the governing equation and the boundary conditions on each element.

The FDM provides an approximated solution of the governing partial differential equation using the Taylor Series approximation. In other words, the partial differential equations are converted to a system of finite-difference equations or linear algebraic equations. The accuracy of the solution depends on the order expressed in Taylor's Series expansion. In summary, the error of the solution decreases as the order of the polynomial function increases.

The LPM simplifies the governing equation assuming that the heat transfer in the satellite and its vicinity can be represented as an electrical circuit. This method assumes that there are isothermal elements in the domain the thermal conductivity is considered infinite and therefore the thermal resistance is minimum. The LPM is applicable when the Biot number is less than 0,1.

---

[8] Meseguer, J., Pérez-Grande, I. and Sanz-Andrés, A., Spacecraft thermal control (1st ed.), Woodhead Publishing, 2012.

[9] Bhaskaran, R., A hands-on introduction to engineering simulation, edX, retrieved 5 June 2022, from https://www.edx.org/course/a-hands-on-introduction-to-engineering-simulations, 2022.

Besides the EFM, DFM, and LPM approach it is required additional models to solve the second term of the LHS Eq. 1. The view factor is the fraction of the energy emitted among the satellite's surfaces. The methods to quantify the radiation emitted due to these surfaces are as follows:

- double area summation,
- Nusselt Sphere technique,
- Crossed-String method,
- Monte Carlo ray tracing,
- contour integration and
- Hemicube.

In this work thermal analysis with the Hemicube approach which is based on the Nusselt Sphere method, is performed. Nusselt's analogy assumes that any surface has the same factor, and all these surfaces cover the same area on the hemisphere. The Hemicube approach projects the surfaces onto an imaginary cube that is around the center of the receiving patch. The advantage of the Hemicube method on Nusselt's analogy is that the first technique is based on the Cartesian coordinate system which simplifies the solution.[10] Miao et. al.[11] presents an extended description of the methods to calculate the view factors.

The classification of commercial software available based on the methods used to solve the general Eq. 1 is presented in Fig. 4. The software used in this work was ANSYS which is based on the FEM and the Hemicube approach to calculate the view factors.

5 Lesson Learned on the TCS Design and Analysis Process

The thermal analysis is an iterative process that involves the communication of all respective subsystem teams to get a successful integration and update the TCS design each time that it should require. In general, there are five phases involved in the thermal analysis[12,13]:

- Phase A: Concept and technology development.
- Phase B: Preliminary design and technology completion.
- Phase C: Final design and fabrication.

[10] Kohnke, P., ANSYS Theory Reference-Release 5.6, in Theory Reference (Issue November). ANSYS Inc, 1994.

[11] See footnote 6.

[12] NASA, NPR 7123.1C: NASA Systems Engineering Processes and Requirements, https://nodis3.gsfc.nasa.gov/displayDir.cfm?t=NPR&c=7123&s=1B, February 2020.

[13] ECSS, ECSS-E-HB-31-03A–Thermal analysis handbook. European Cooperation for Space Standardization,. https://ecss.nl/home/ecss-e-hb-31-03a-15november2016/, 15 November 2016.

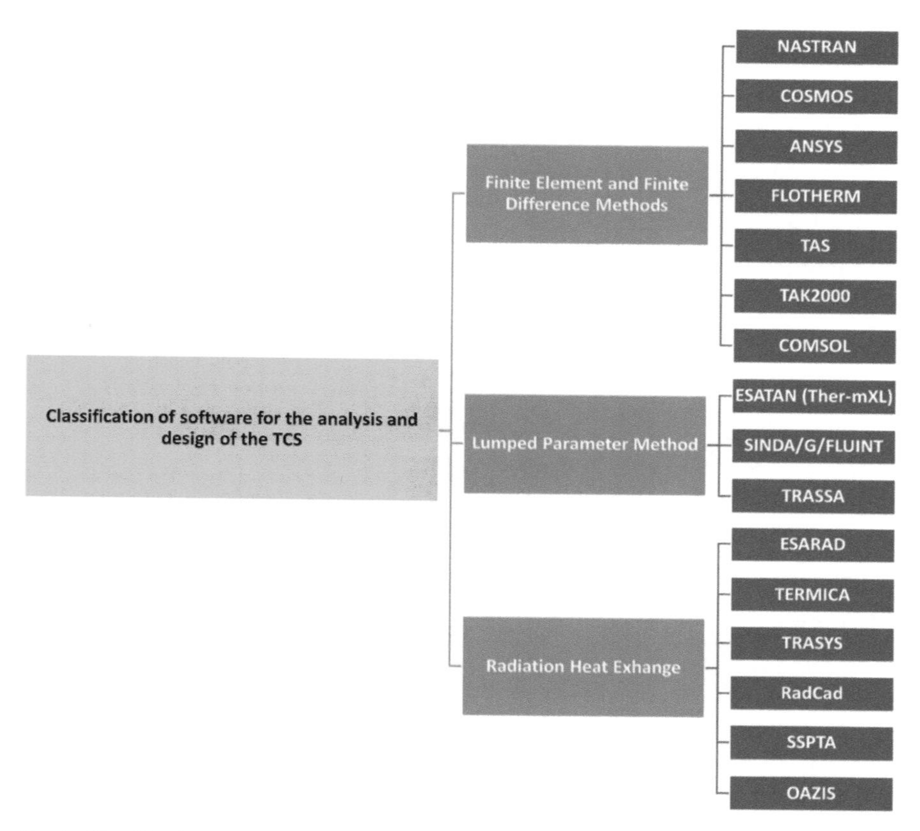

Fig. 4 Classification of software for the analysis and design of the TCS summarized by Baturkin, V., Micro-satellites thermal control—concepts and components, Acta Astronautica, 56(1), pp. 161–170, https://doi.org/10.1016/j.actaastro.2004.09.003, 2005

- Phase D: System assembly, integration, test and launch.
- Phase E: Operation and sustainment.

Figure 5 shows the process and the stages performed for thermal analysis of the K'OTO satellite. At this point, the development of the K'OTO CubeSat is at Phase C: Detailed definition (final design and fabrication).

Phase A involves the review of the systems requirements such as the definition of the TCS concept and the identification of the requirements of the thermal analysis.

Phase B involves the calibration of the base case reviewing the material properties, the contacts among components, the optical properties, and the analysis of worst-case scenarios such as the hot case and the cold case.

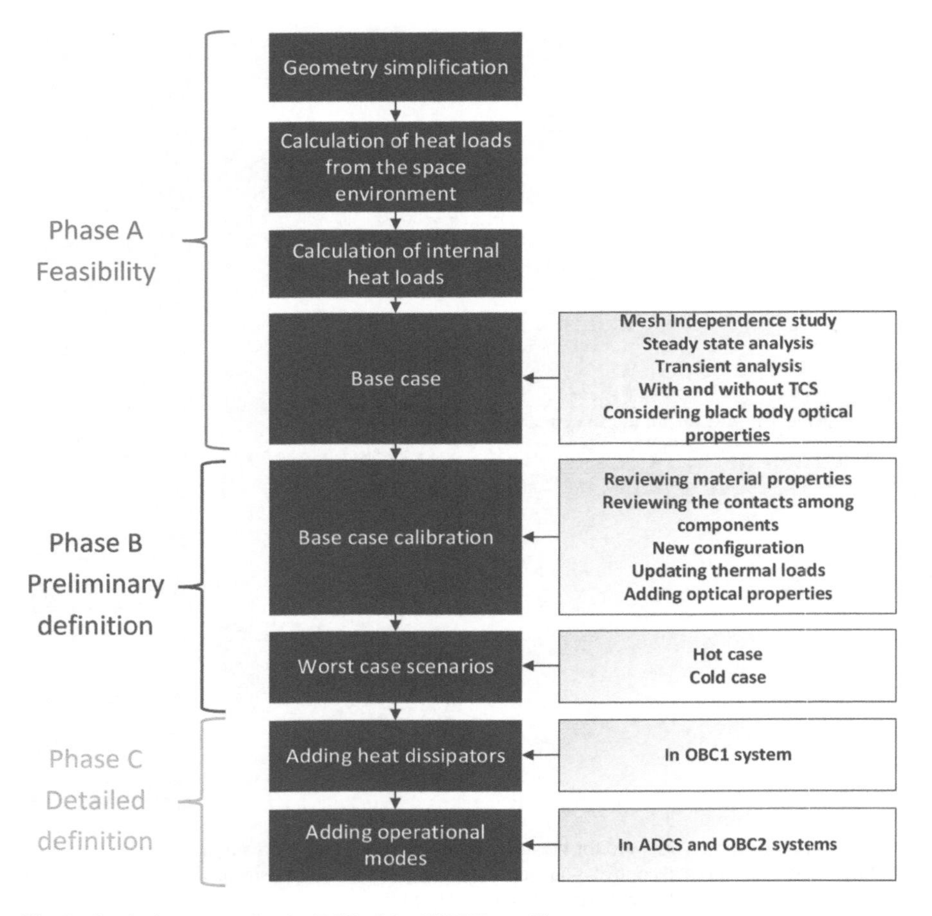

Fig. 5 Analysis process for the TCS of the K'OTO satellite

Phase C is the detailed definition of the K'OTO satellite including a heat dissipator and the operational modes of the critical systems which in Phase B do not meet the thermal requirements.

Phase D and Phase E are under preparation because the certification test and launching of the K'OTO satellite are planned to be launched in 2023.

During Phase A the following substeps were identified[14]:

- simplification of the satellite's geometry,
- review of contacts,

[14] Gaviria-Arcila, D., Ferrer-Perez, J. A., Chavez-Moreno, R. G., Romo-Fuentes, C. and Ramírez-Aguilar, J. A., Análisis numérico para el sistema térmico del nanosatélite K'OTO. Memorias del XXVII Congreso Internacional Anual de la SOMIM, 22 September 2021.

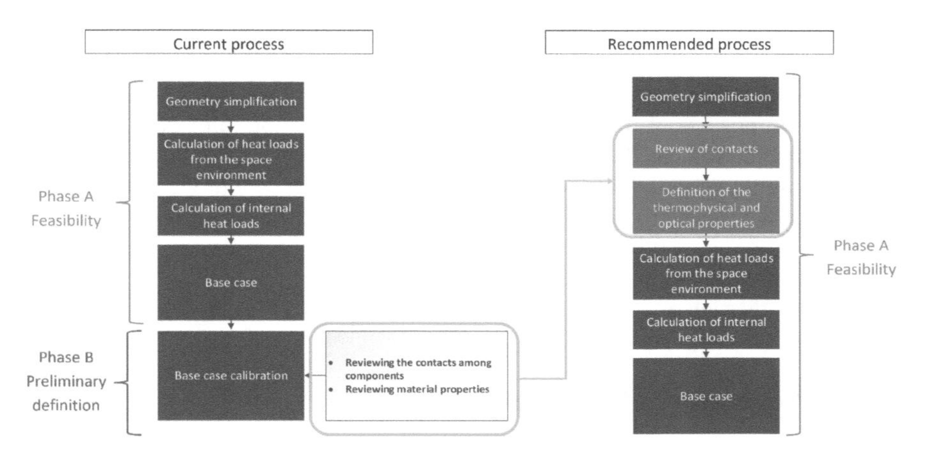

Fig. 6 Recommended process for the Analysis of the TCS based on lessons learned. On the left Current process, on the right Recommended process

- definition of the thermophysical and optical properties,
- preparation of thermal loads,

  o calculation of heat loads from the space environment,
  o calculation of internal heat loads and

- set up and modeling of the base case.

The lesson learned to highlight from this process is that a review of contacts along with the satellite geometry and the definition of the thermophysical and optical properties should be performed before starting the numerical solution.

Figure 6 shows the current process of Phase A and the recommended process to perform the thermal analysis. In Fig. 6 the current process reviews contacts and definition of thermophysical and optical properties until Phase B within the Base case calibration. However, based on the experience of the thermal analysis of the K'OTO satellite, the recommended process proposes to perform the review of contacts and the definition of thermophysical and optical properties after the geometry simplification i.e. during Phase A. The advantage of the recommended process will allow saving time on the calibration process during Phase B and avoid errors during the numerical solution.

5.1 Simplification of the Satellite'S Geometry

The simplification of the satellite's geometry is the first step in the thermal analysis for the TCS. This step allows for saving computational resources in meshing and simulation and it is relevant because the analysis is focused on finding regions at the risk of damage or malfunction due to not satisfying the temperature specifications. The thermal analysis does not need the level of details on the geometry as it is in vibration or structural analyses.

The simplification of the geometry is based on the remotion of details such as chamfers and complicated geometries, and the remotion of connectors and fasteners. The simplified model just includes the boards and chips that we assumed had high levels of heat dissipation such as the processors as shown in Fig. 7.

Recommendation
Remove details such as complicated geometries. Remove connectors and fasteners.

5.2 Preparation of Thermal Loads

5.2.1 Calculation of Heat Loads from the Space Environment

The calculation of heat loads from the space environment was obtained from the Systems Tool Kit (STK 12) software from the Keplerian elements of the International Space Station and post-processed to calculate the external heat loads as shown in Fig. 8.

Recommendation
Consider the external loads from the worst-case scenarios, namely the hottest and the coldest day of the year.

5.2.2 Calculation of Internal Heat Loads

The calculation of internal heat loads was obtained from the theory of the two-resistor compact thermal model which simplifies the model to three nodes that are connected by two resistors. These three nodes consist of the board node, the junction node, and the junction to case.[15] This calculation was performed with the COTS which has data available and provided by the supplier.

Recommendation
Estimate the internal loads from the two-resistors thermal model.[16]

5.3 Set up and Modeling of the Base Case

The thermal analysis for the TCS of the K'OTO was performed through the FEM with the Hemicube approach using the ANSYS software and its thermal module version 2021 R1.4. The main goal of this analysis is to find the distribution of

[15] JEDEC, JESD15-3, Two-Resistor Compact Thermal Model Guideline, https://www.jedec.org/standards-documents, issue July 2008.
[16] See footnote 13.

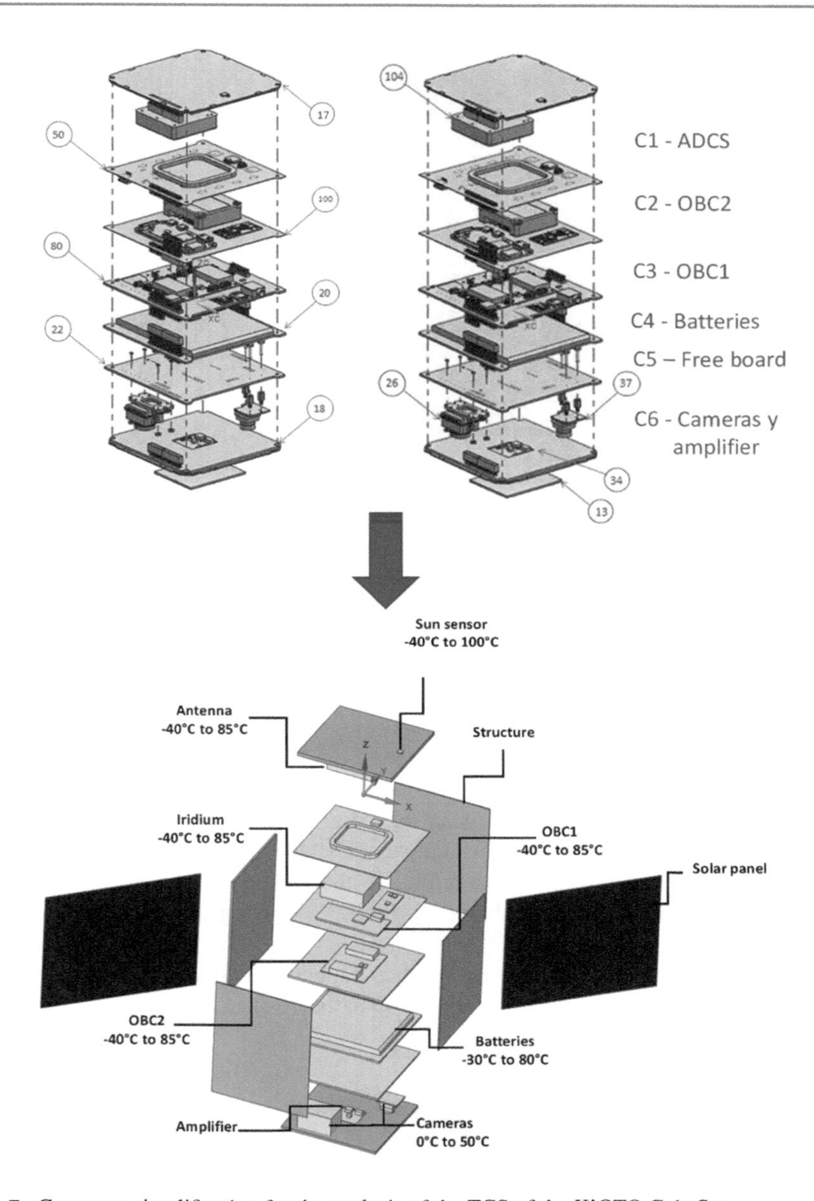

Fig. 7 Geometry simplification for the analysis of the TCS of the K'OTO CubeSat

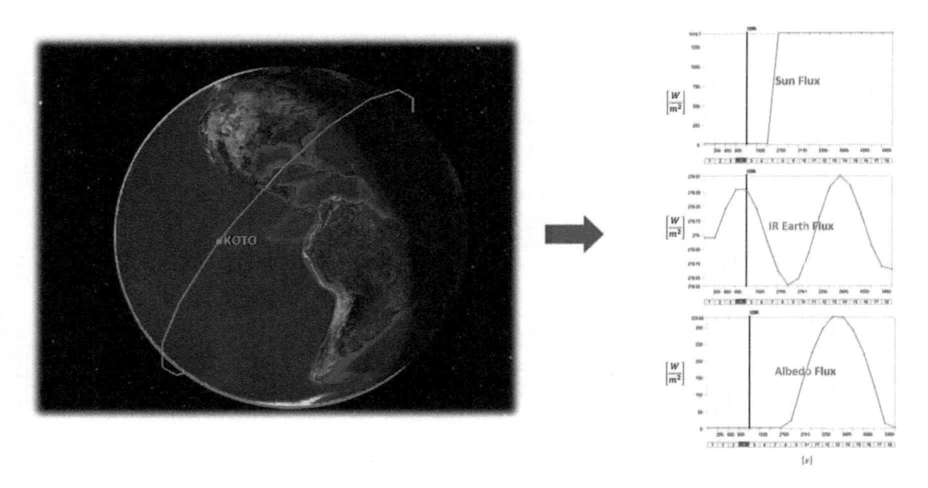

Fig. 8 External heat loads for the thermal analysis of K'OTO CubeSat

temperatures in the satellite and determine the components which are at the risk of under passing or overpassing the operational limits of temperature required to have an optimal function.

5.3.1 Mesh

The setup and modeling of the base case used the FEM to predict the temperature distribution of the satellite including the internal and external heat loads as a function of time. Figure 9 shows the mesh used for the discretization of the K'OTO CubeSat geometry and its thermal analysis. The mesh has three zones of mesh sizing, where sizing is reduced according to the regions which need to increase the accuracy of predictions as shown in Table 1. The regions of interest are the chips, processors, and cameras.

The setup of the case involves a mesh independence study to verify the convergence of the results and to determine that the results are independent of the

Fig. 9 Mesh used for the thermal analysis of K'OTO CubeSat

Table 1 Mesh sizing

Region	Sizing [m]					
	Mesh 1	Mesh 2	Mesh 3	Mesh 4	Mesh 5	Mesh 6
Chips and processors	3,16E-03	1,58E-03	1,44E-03	1,31E-03	1,19E-03	1,08E-03
Board and structure	4,823E-03	2,41E-03	2,19E-03	1,99E-03	1,81E-03	1,65E-03
Cameras	2,77E-03	1,39E-03	1,26E-03	1,15E-03	1,04E-03	9,47E-04

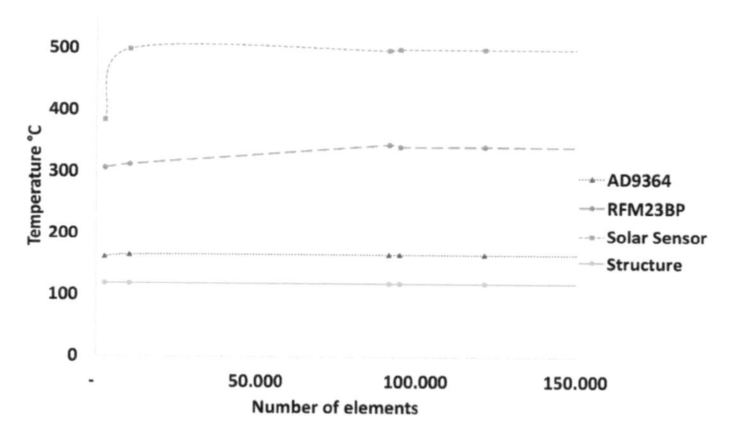

Fig. 10 Mesh independence study

element size. Figure 10 shows the mesh independency study. Three components and the satellite's structure were monitored and five meshes (see Table 1) were proposed to verify the independency of results with the element size. Mesh 3 was the optimal mesh size because from this it is noticed that the variation of temperature is minimum. Mesh 3 is highlighted in yellow in Table 1 which shows the sizing used to discretize the three regions of interest.

Recommendation
Get a base case and results with a coarse mesh as a starting point. Once you get the first results perform a mesh independence study.

5.3.2 Base Case

As part of the setup and of the base case was implemented different cases with and without TCS to observe the temperature variations, as well as simulate a steady-state analysis and a transient analysis where it was noticed that the transient case analysis can include the temperature variation due to the eclipses presented during the orbit of the satellite. Table 2 shows the cases presented and discussed in this work with the lesson learned in each one.

Table 2 Cases discussed for the base case

Case	State	Details
1	Steady state	One cycle (one orbit)
2	Transient	One cycle (one orbit)
3	Transient	One cycle (one orbit)
4	Transient	Five cycles (five orbits)

Steady-State Case Versus Transient Case

In this section, the results of two cases are presented: Case 1: the Steady state case and Case 2: Transient state case. On the one hand, Case 1 was analyzed with ANSYS Steady-State Thermal module. It considers constant boundary conditions namely the external and internal heat loads. On the other hand, Case 2 was performed with ANSYS Transient Thermal Module, for this case the boundary conditions for the external loads were considered a function of time. However, the internal loads were considered constant. The external loads are periodic loads due to the eclipses that the satellite presents in its orbital path. In both cases, all satellite components and the satellite's structure itself are considered to radiate into the environment with a temperature of 18 °C and an emissivity of 1. The initial temperature is uniform at 22 °C. For the transitory case, one cycle is considered, that is, one return to the Earth. The internal heat loads considered for both iterations are shown in Table 3. Both cases assume one cycle (one orbit).

Figure 11 shows the boundary conditions applied to the satellite where Fig. 11a shows the heat loads from the external environment and Fig. 11b shows the internal heat dissipation from the satellite's components.

Figure 12a and b show the temperature gradients of the entire satellite for the stationary case and the transitory case of a cycle, respectively. It can be seen that in Fig. 12a and b the maximum temperature is found in the solar sensor. However, the steady state case presents a temperature of 36% higher than the transient state. This is because the transient state considers a period of low temperature, hence it represents what is happening in the satellite and its orbital movement.

The maximum value reached in the solar sensor exceeds the maximum operating limit, which is 100 °C. For the base case, the material used for the faces of the satellite was considered AL6061-T6. However, the material used should be the FR4 which is one of the reasons that the temperature of the solar sensor reaches this value. Another reason was that the geometry of the solar sensor shows an overlap with the face of the satellite. Therefore, the calibrated model presented in Sect. 5.4 includes these two modifications, namely, changing of material and verifying the physical contact regions between surfaces.

Altogether, two cases were modeled: a steady state case and a transient state case. Where it was observed that the steady state case can be used as a first approximation to estimate the temperature of the satellite. However, it is recommended to use the results of the transient state case to obtain a model closer to reality. From the numerical analysis, it was possible to observe that the component with the highest temperature is the solar sensor of the upper part of the satellite, the solar

Table 3 Summary of boundary conditions and considerations taken for the thermal analysis

Heat Flux [W/m^2]	
Sun	1.414
Earth IR	275
Albedo	805,98
Heat flow [W]	
ADCS-STM32F103C8T6	0,363
ADCS-LTC24971UHFPBF	1,76
ADCS-BD6212HFP-TR	0,832
ADCS-MT01	2,66
OBC2-Iridium	3,75
OBC2-Raspberry	0,875
OBC2-RFM23BP	0,911
ICEPS-AD9364 RFIC	0,774
ICEPS-XILINX ZYINQ SoC 7010-2i	0,8
ICEPS-ISSI-is43tr16256al-125kbl	0,33
14 EXA Pegasus Class Battery	11,1
Solar sensor	0,98
Radiation	
Emissivity	1
Ambient temperature	18 °C

sensor shows a maximum temperature in all cases where it is recommended to calibrate the model in such a way as to ensure that the conditions being analyzed are close to the real conditions in which the satellite is found.

Recommendation
Run a steady state case for a first estimation and a transient state case to get a close insight into the reality.

5.4 Base Case Calibration

5.4.1 Review of Contacts

In this section, we do reference contacts defining them as the mechanical interfaces between components. The thermal contacts should be well-established along with the model because, from these, the model can estimate the phenomenon of heat transfer by conduction among the components and the structure of the satellite. The thermal contacts between components depend physically on the rugosity of materials, the adherent, the pressure distribution within contacts, the contact

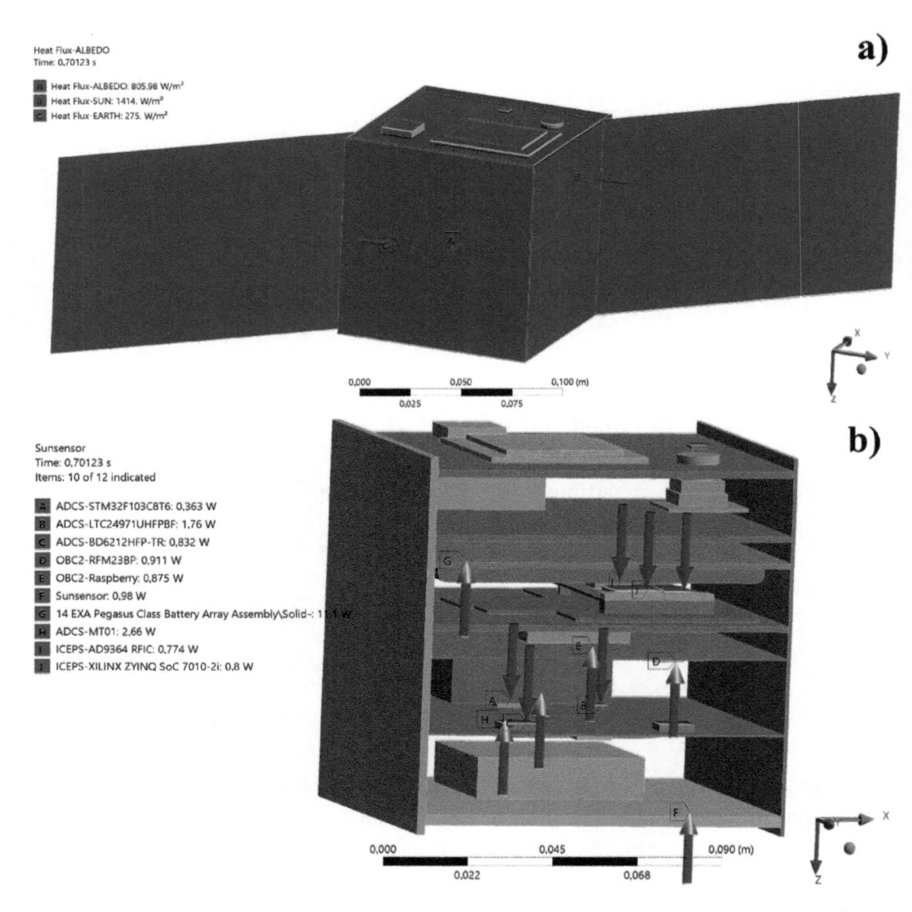

Fig. 11 Boundary conditions. **a** External heat loads, **b** Internal heat loads

area, the geometry, and the machining imperfections of the parts, among others.[17] The software interprets a contact condition when two separated bodies touch one another i.e. they are mutually tangent.[18] To simplify the model, we consider the thermal contacts between components as surface-to-surface contacts assuming that are bonded. The bonded contact is considered lineal contact without separation between faces. The bonded contact considers that there is no sliding between faces assuming that the surfaces are as glued ignoring any friction or

[17] D.G. Gilmore and M. Donabedian, Spacecraft thermal control handbook, 2nd ed., Aerospace Press, El Segundo, Calif, 2002.

[18] ANSYS, ANSYS Mechanical User's Guide - Release 2021 R2, https://ansyshelp.ansys.com/account/secured?returnurl=/Views/Secured/corp/v212/en/wb_sim/ds_contact_overview.html, July 2021.

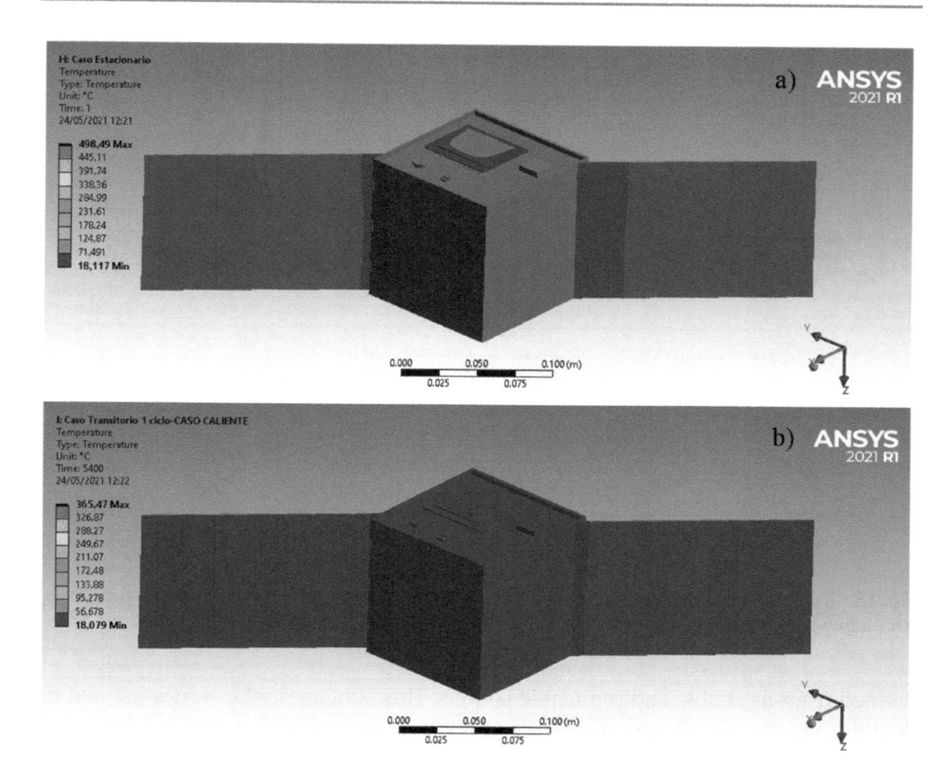

Fig. 12 Temperature gradients of the entire satellite. **a** Steady state Case; **b** Transient case (one cycle, hot case)

rugosity at the surfaces. This type of contact allows a linear solution since the contact area is not variable[18]. We use the automatic algorithm which is program-controlled to compute the thermal interface between surfaces. The tolerance used was 1.1405e-003 m to avoid any interpenetration or overlapping between bodies.

It is very important to review the contacts among the bodies to verify that there are no overlapping contact regions and there is surface-to-surface contact. Avoiding this recommendation may cause a wrong estimation of the temperature of components because the heat transfer by conduction may not be to represent the physical phenomena.

Figure 13 shows the two bodies before and after the review and definition of contacts. At the top of Fig. 13 we can observe the overlapping bodies, and at the bottom it is shown that the contact is corrected showing the same bodies with a surface-to-surface contact.

5.4.2 Definition of the Thermophysical and Optical Properties

During the thermal analysis of K'OTO, it was noticed that before starting the solution of a case it is important that the thermophysical and optical properties are well

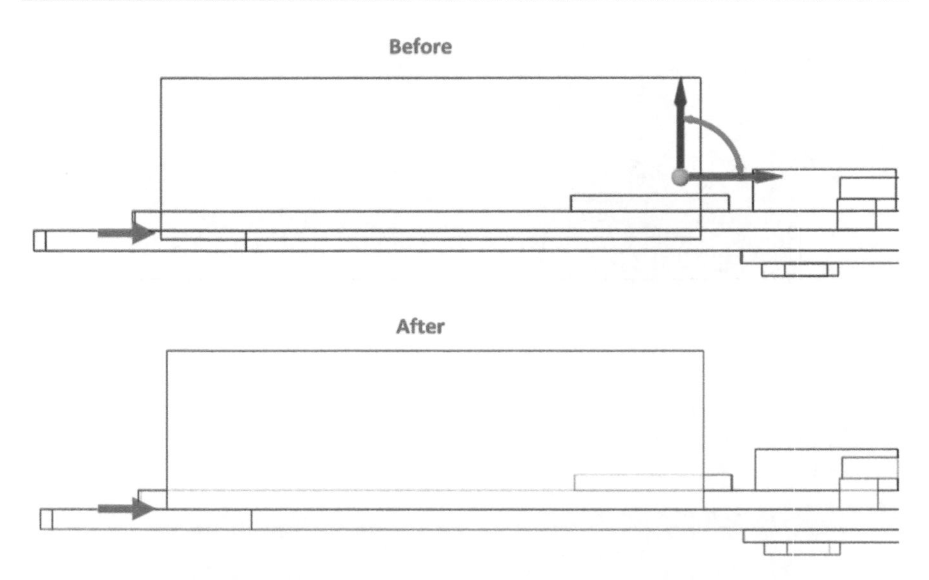

Fig. 13 Representation of definition of contacts. Top: two overlapping bodies, Bottom: two bodies with a bonded contact

defined from a reliable and replicable source. This is because the values defined by default may not be the values that are applicable for materials previously used for the analysis of satellites. One example of this is the thermal conductivity of FR4 used for COTS components which may vary according to the cupper layers used. The results of the base case show that the temperature of the solar sensor overpass the operational temperature range and is even above the soldering temperature, i.e. above 260 °C. The main reasons for this issue were the definition of the thermal conductivity of FR4 and the contact between the solar sensor and board which showed a gap and the heat transfer by conduction was not calculated properly.

Repeated experiments were performed to observe the sensibility of results due to the material's properties. One experiment consists of transmitting and receiving data using the transceiver module RFM25BP and measuring the temperature with an IR camera every five seconds. Figure 14 shows the results of this experiment where the temperature reaches the steady-state at 33 s with a temperature of 63 °C. The evaluation of the sensibility of material's properties to the temperature estimation of the transceiver module was done in four cases. The main parameter to evaluate was the thermal conductivity of FR4. The considerations used in the numerical analysis are presented in Table 4. These considerations were common in all cases.

Table 5 shows the cases to evaluate the sensibility of material's properties on the temperature of the transceiver module RFM23BP. Four cases were simulated varying the thermal conductivity k of FR4. The Cases A, B, and C consider a Specific Heat Capacity Cp of 600 [J/kgK] and Case D considers a Cp of 300

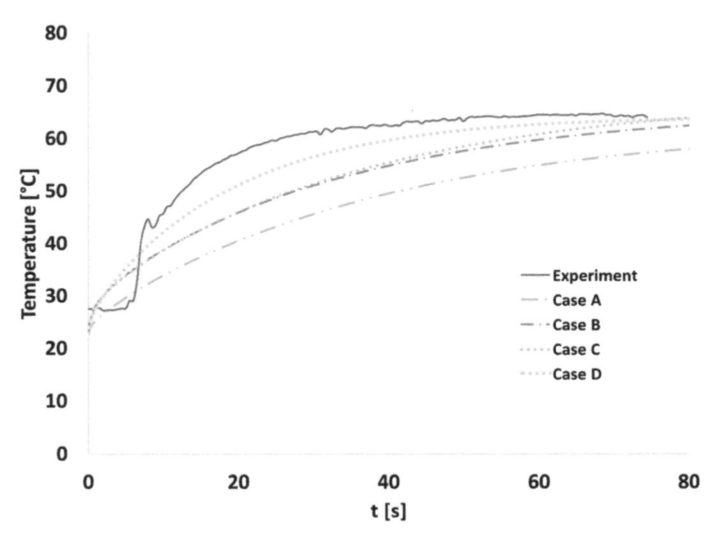

Fig. 14 Evaluation of the sensibility of the evolution of temperature on the transceiver module (RFM23BP) due to the material properties of FR4

Table 4 Considerations used for the numerical analysis to evaluate the sensibility of material's properties on the temperature of the transceiver module RFM23BP

Power dissipation of chip [W]	0,6
Convective coefficient [W/m^{2}K]	16,45
Emissivity	0,95

[J/kgK]. The four cases were compared against the experimental results from the Root Mean Squared Error (RMSE) which is shown as well in Table 5.

Figure 14 shows the comparison of experimental data versus the numerical results. The four cases follow the tendency of the experimental data. On the one hand Case A present an RMSE of 8,6 which is the highest of all cases. On the other hand, D presents an RMSE of 4,7 which is the smallest error compared to the other three cases (see Table 5). The cases A, B, and C show that the temperature is

Table 5 Cases simulated to evaluate the sensibility of material's properties of FR4 on the temperature of the transceiver module RFM23BP

Case	k [W/m^{2}k]	Cp [J/kgK]	RMSE
A	26,55	600	8,6
B	17,7	600	7,3
C	8,85	600	5,8
D	17,7	300	4,7

sensible for both parameters namely, the thermal conductivity and the specific heat capacity of the FR4. Where it is observed that by varying these two parameters, the temperature estimate for case D is the closest estimate to the experimental data. More research is being done on the sensibility of the predictions of satellite's temperature due to the material's properties; however, this is a general overview of why it is important to define the properties of materials well from the beginning of the analysis.

Recommendation
Define the thermophysical and optical properties of materials, from previous reports or reliable literature, at the beginning of the thermal analysis process.

5.4.3 One Orbit Versus Five Orbits

This section evaluates the hot scenario for Case 3 with one cycle or orbit and Case 4 with five cycles or five orbits (according Table 2). The main purpose of this section is to review whether the simulation of a case with five cycles is needed to represent the thermal performance from the orbital movement at its effects on the satellite. Specifically, it is to observe the variation of the thermal performance of the satellite including a case with one orbit and a case with five orbits.

Figure 15 shows the comparison of Cases 3 and 4, where it is observed that the graph of both cases follows the same trend. The percentage of error between both cases is less than 1% as observed in Fig. 16.

The error was calculated as follows,

$$\text{Error} = \left| \frac{v_A - v_E}{v_E} \right| \cdot 100\% \tag{3}$$

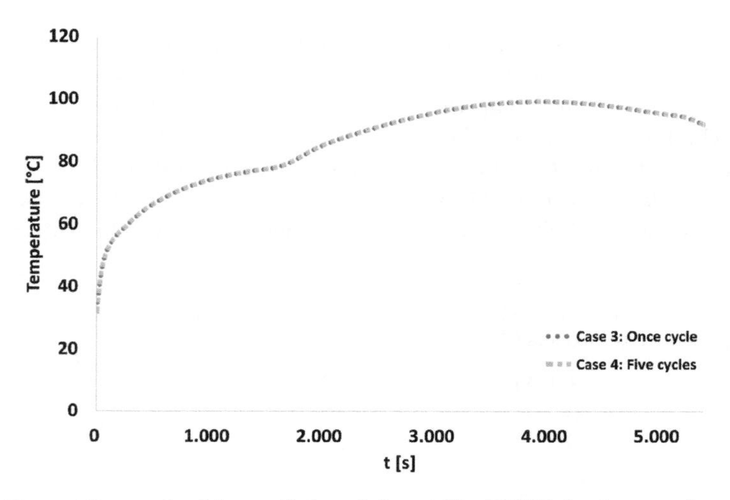

Fig. 15 Comparative graph of the prediction of the satellite K'OTO for the worst hot case scenario. Case 3 with one cycle (one orbit) and Case 4 with 5 cycles (five orbits)

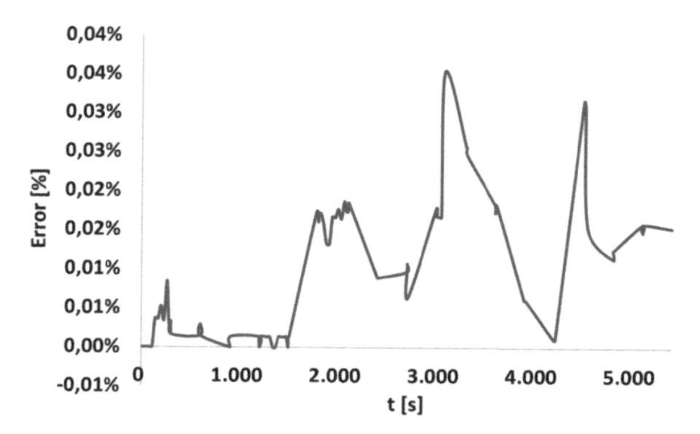

Fig. 16 Percentage of error calculated from the temperature predictions of K'OTO from a worst hot case scenario between case 3 with one cycle (one orbit) and case 4 with five cycles (five orbits)

where v_A is the actual value and v_E is the expected value.

The graph in Fig. 15 shows the 5.400 s of the cycle. However, Fig. 17 shows the temperature variation of the K'OTO satellite for the transient state case including five cycles where the maximum temperature is seen in the green curve, the average temperature in the blue curve, and the minimum temperature in the red curve. From here it can be seen that the maximum point of Case 4 reaches a temperature value of 99,184 °C which differs from Case 3 with one cycle by less than 1 °C since for Case 3 the maximum temperature value is 99,074 °C.

The comparison of both cases showed that the error percentage is less than 1%, so it is concluded that Case 3 allows obtaining a prediction with an error percentage of 0,04% compared to Case 4. Furthermore, Case 3 with one cycle uses less computational resources, that is, less simulation time. Therefore, it is recommended to carry out the numerical simulations for one orbit (one cycle) (Fig. 15), since the results are just as reliable as for five orbits (five cycles) (Fig. 17).

Recommendation

A transient case with one orbit will provide similar results to a case with five orbits but with less computational time.

5.4.4 Operational Modes

The components of the ADCS system of the K'OTO satellite operate with six operational modes as follows:

- Reset: the global variables and the registered values are restarted.
- Idle: there is a low consumption mode in the subsystem components.
- Housekeeping (HK): the status of peripheral components and sensors is checked and stored.
- Slave: there are periodical measurements of the data collected by the available sensors and peripherical components and this information is stored in a register.

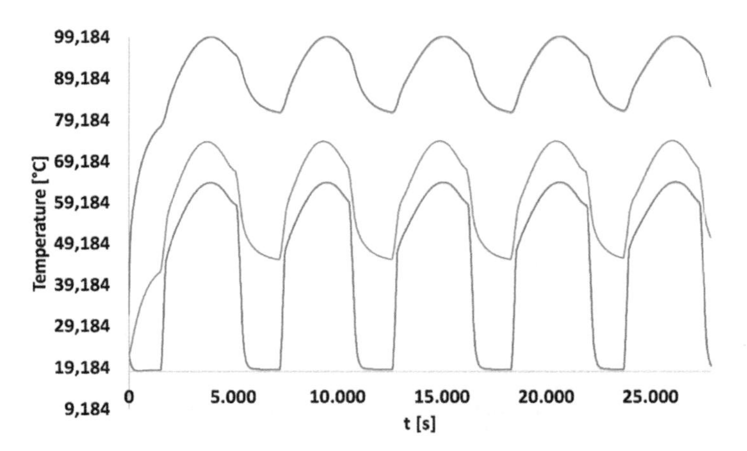

Fig. 17 Temperature variation on the K'OTO satellite for Case 4 with five cycles (five orbits)

- Detumbling: measurements are made of the data collected by the sensors; the data is processed and the magnetorquers are activated at an indefinite sampling time.
- Pointing: same that the detumbling mode. Here the data processed is used to determine the orientation and pointing of the satellite's Z-axis towards Nadir.

At this point, the analysis presented in this section is focused on the LTC2497 component which was one of the components that does not meet the operational range of temperature. The operational mode taken as a reference for the LTC2497 is the pointing mode, where the component turns on cyclically for 0,75 s and turns off for 0,25 s as shown in Fig. 18.

For this analysis, it is proposed that the cyclic operation of the pointing operational mode is represented by an equivalent value. This equivalent value is estimated through the mean value theorem for integrals which ensures that a continuous function on a closed interval reaches its average value at least at one point. The graphical representation of this theorem is the area under the graph of an interval (i.e. 0 W to 1,176 W) is equal to the area of a rectangle whose base is the length

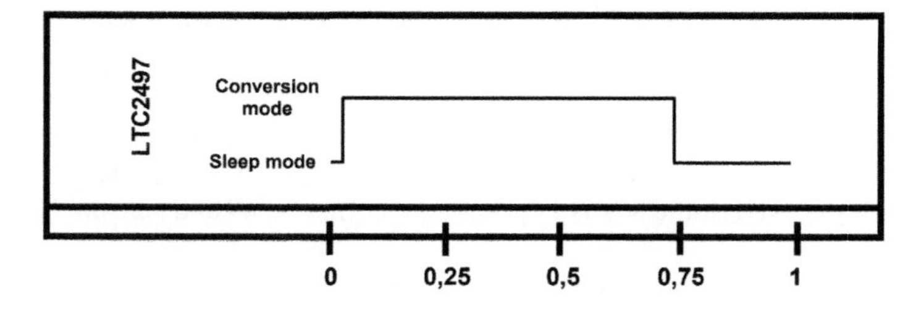

Fig. 18 Diagram of state of the pointing mode for the LTC2497 component

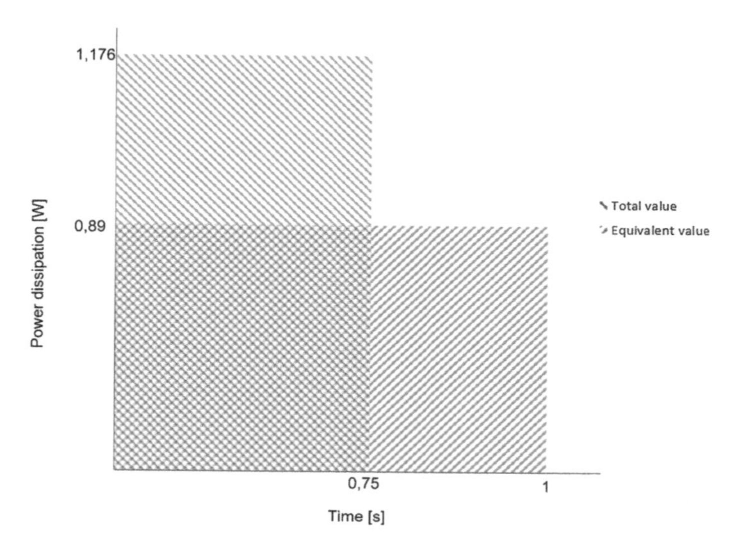

Fig. 19 Graphical representation of the estimated equivalent (mean) value for the LT2497 component

of the interval and height is the value of the integrand at some point in the interval (0,89 W)[19] as shown in Fig. 19. Therefore, the average value along the cycle is 0,89 W which is the equivalent value used to represent the operational mode of the component LTC2497.

This is detailed as follows,

The area of the blue rectangle A_{tv} shown in Fig. 19 is given by

$$A_{tv} = b \times h = 0{,}75 \times 1{,}176 = 0{,}89 \tag{4}$$

where b is the base of the rectangle and h is the height of the rectangle.

Therefore, the mean value is given by,

$$A_{ev} = b \times h = 1 \times h = 0{,}89 \therefore h = 0{,}89 \tag{5}$$

where A_{ev} is the area under the curve of the mean or equivalent value.

The same procedure was implemented for the components of the On-Board Computer (OBC) which similarly to the components of the ADCS were at risk due to overpassing the operational limits of temperature. Table 6 presents the total value and the mean value of the power dissipation in Watts. It was assumed that

[19] Meade Douglas B., Mean Value Theorem for Integrals, Department of Mathematics, University of South Carolina, retrieved 05 May 2022, https://people.math.sc.edu/meade/Bb-CalcI-WMI/Unit5/HTML-GIF/MVTIntegral.html#:~:text=The%20Mean%20Value%20Theorem%20for,average%20value%20on%20the%20interval, 2013.

Table 6 Total and equivalent values for the power dissipation of the ADCS and OBC components represent the housekeeping operational mode

System	Component	Power dissipation [W]	
		Total value	Equivalent (Mean) value
ADCS	LTC2497 Analog–digital converter	1,176	0,89
	MT01 Magnetorquer	2,66	0,665
OBC2	RFM23BP Transceiver module	3,75	0,68
	MIKROE-1204 Interface development tool	0,8	0,61

the housekeeping operational mode of the OBC works with the same cycle as the ADCS system, namely, the components turn on for 0,75 s and turn off for 0,25 s.

Once implemented the operational modes for the ADCS and OBC components, the temperature of these components meets the temperature requirements and there was no need to propose an active TCS.

Recommendation
Implement the operational modes of the satellite components using the equivalent or mean value for the power dissipation.

5.5 Conclusions

This work presented the lessons learned for the thermal analysis implemented on the K'OTO satellite. The goal of this work is to present step by step all the main challenges faced during the thermal analysis of a small satellite and based on this reduce the learning curve of the thermal analysis of future missions using computation tools based on FEA. Additionally, along with the manuscript is presented the advantages of the numerical analysis, and the numerical methods used to determine the temperature distribution required to design the TCS of small satellites. During the thermal analysis of K'OTO, there were identified different challenges to perform the numerical modeling and predict the temperature of the satellite. The challenges presented were the definition of thermal and optical material properties, the simplification of geometry, the definition of contacts among components, the preparation of thermal loads, the change in temperature over time, and the effects of the internal component distribution on the variation of temperature. One of the lessons learned during the thermal analysis of K'OTO is that the correct selection of the material properties is highly relevant to have a reliable prediction of the design of the thermal control system as well as the definition of contacts among components. Based on these methods is presented the classification of the software used for the thermal analysis and design, and more important, enclosed hints are enlisted to emphasize technical considerations to generate robust and well-based thermal numerical simulations. With this guide, it will be possible to design more reliable TCS for CubeSats.

Acknowledgments The authors are very grateful to the Mexican National Council for Science and Technology (CONACYT) and its postdoctoral fellowships for Mexico 2020 and 2021 for their valuable financial support. Moreover, the authors thank Project UNAM PAPIIT IT101522 for its support. We want to express our gratitude to "Grupo SSC" for providing support and the ANSYS license for developing this project.

Dr. Dafne Gaviria-Arcila is a Research Fellow at the Advanced Technology Unit of the National Autonomous University of Mexico. She has been involved in thermal and fluids analysis for more than 7 years. She was awarded 2020 her Ph.D. from the University of Nottingham in England. As part of her Ph.D. project, she was a recipient of the ZONTA International, Amelia Earhart Fellowship in 2017. Because of this recognition, in 2018 the Honorable Mention of Carlos Fuentes Award from The Mexican Embassy in the United Kingdom. She has participated in different activities to influence women to study STEM careers such as mentoring, inspiring, and empowering them. Her story has been featured in the 100 extraordinary Mexicans edition of the book titled Good Night Stories for Rebel Girls.

Dr. Jorge A. Ferrer-Pérez is a professor at the National University Autonomous of Mexico-School of Engineering. He received his Ph.D. in Aerospace and Mechanical Engineering from the University of Notre Dame, South Bend in the United States. He is part of the Aerospace Engineering Department and is responsible for the Space Propulsion and Thermo-vacuum lab. This facility belongs to the Space and Automotive Engineering National Laboratory located at Juriquilla. His current research areas are nano-heat transfer in solid-state devices, thermal control, space propulsion, small satellites, and the development of space technology.

Dr. Carlos Romo-Fuentes is an associate professor at the National University Autonomous of Mexico School of Engineering. He received his Ph.D. in Technical Sciences in the Design of Space Systems considering electromagnetic compatibility criteria from the Aviation Institute of Moscow, Russia. He is part of the Aerospace Engineering Department and is responsible for the Electromagnetic Compatibility Laboratory. His current research areas are electromagnetic compatibility, certification tests, space systems, and space technology development. Likewise, is the technical responsible for the Space Science and Technology Theme Network from the National Council of Science and Technology from the Government of Mexico.

Dr. Rafael G. Chávez-Moreno is an assistant professor at the National University Autonomous of Mexico-School of Engineering. He received his Ph.D. in Mechanical Engineering from the School of Engineering-UNAM. He is part of the Aerospace Engineering Department and is responsible for the Model-Based on Design lab which belongs to the Space and Automotive Engineering National Laboratory located at Juriquilla. He is an active member of the Mexican Society of Mechanical Engineering and the Space Science and Technology Network. His current research areas include space systems, embedded systems, and control systems.

Dr. Jose Alberto Ramírez-Aguilar is an assistant professor at the National Autonomous University of Mexico-School of Engineering. He received his Ph.D. in Technical Sciences in Radio receivers and microsatellites from the Moscow Aviation Institute - MAI, Russian Federation. He is the head of the Aerospace Engineering Department. His current research areas are Radio Frequency and microwave Systems, GNSS, Antennas, TT&C, Nano, and Microsatellites. Likewise, in 2020 was selected for the first Latin American manned space mission ESAA-01EX SOMINUS AD ASTRA.

Dr. Marcelo López-Parra has a Doctorate Degree in Advanced Manufacturing granted by the Cranfield Institute of Technology, England. He has taught at bachelor's and master's levels for 40 years, having supervised 29 bachelor's thesis projects, 33 master's theses, and 9 doctoral theses. His production includes 65 collaboration joint projects with the industry, 152 publications in magazines and conference proceedings, and six registered patents. He was awarded the Technology Development Prize in 2012, State of Nuevo León, México. Dr. López-Parra currently holds top-level 3 within the National Research System. He was a member of the evaluation commission of the National Research System (S.N.I.), engineering, (2003–07), a member of the technology evaluation subcommittee in the same S.N.I. (2011–15), and a member of the University UNAM Council representing the campus of C.U. (2012–2016). Since the year 2018, he has been the Head of the Advanced Technology Unit (UAT) at UNAM, Querétaro's Campus.

K'oto Project a Cubesat Design: Methodology and Development

Rafael G. Chávez-Moreno, Jorge A. Ferrer-Pérez, Carlos Romo-Fuentes, José A. Ramírez-Aguilar, Sergio Ríos-Rabadán, María G. Ortega-Ontiveros, Xochitl Silvestre-Gutiérrez, Eduardo Muñoz-Arredondo, Saúl Zamora-Hernández, Edgar I. Chávez-Aparicio, Saúl O. Pérez-Elizondo, Bryanda G. Reyes-Tesillo, and Dafne Gaviria-Arcila

Abstract

The aerospace industry has grown significantly in Mexico over the last ten years, and it is a sector in constant evolution. However, when talking about

R. G. Chávez-Moreno (✉) · J. A. Ferrer-Pérez · C. Romo-Fuentes · J. A. Ramírez-Aguilar · S. Ríos-Rabadán · M. G. Ortega-Ontiveros
Advanced Technology Unit, UAT/FI/UNAM, Juriquilla Querétaro, México
e-mail: rchavez@comunidad.unam.mx

J. A. Ferrer-Pérez
e-mail: ferrerp@unam.mx

C. Romo-Fuentes
e-mail: carlosrf@unam.mx

J. A. Ramírez-Aguilar
e-mail: albert09@unam.mx

M. G. Ortega-Ontiveros
e-mail: maria.guadalupe.ortega@comunidad.unam.mx

X. Silvestre-Gutiérrez
Durango Institute of Technology, Durango, Durango, México
e-mail: 16041251@itdurango.edu.mx

E. Muñoz-Arredondo
Querétaro Institute of Technology, Querétaro, México
e-mail: 116140994@queretaro.tecnm.mx

S. Zamora-Hernández
Autonomous University of Querétaro, Querétaro, México
e-mail: szamora06@alumnos.uaq.mx

E. I. Chávez-Aparicio · S. O. Pérez-Elizondo · B. G. Reyes-Tesillo
Applied Physics and Advanced Technology Center, UNAM, Juriquilla Querétaro, México
e-mail: echavezaparicio@comunidad.unam.mx

A. Froehlich (ed.), *Space Fostering Latin American Societies*, Southern Space Studies,
https://doi.org/10.1007/978-3-031-20675-7_8

the aerospace industry, it references the aeronautical industry, leaving aside the space industry until a couple of years ago. Recently, significant advances have emerged in the Latin-American space sector because the needs and technologies have accelerated in this industry, fostering the development of a new space age. This has generated significant technological and industrial development opportunities in Latin American countries. This document presents a brief CubeSat design under Mexico perspective, covering the methodology and development taking the K'OTO project as a study case, which consists of the development of a Mexican nanosatellite based on the CubeSat 1U standard, whose mission is the remote perception of the Mexican territory through images in the visible spectrum, being a technological demonstrator. The purpose of the K'OTO project is to encourage the development of the space industry in Mexico, having as priorities the generation of human resources and the development of national technology, as well as awakening interest in the young public about this sector, achieving the leap technology from Mexico into space. Within this document, the management and engineering challenges of space systems applicable to the development of the K'OTO project are addressed, making NASA's NPR 7120.8 a reference but making a tropicalization of it, which is adapted to the conditions of Mexico. Subsequently, each subsystem that makes up the nanosatellite is presented in detail from its definition to its design. Finally, major lessons learned are presented from the development of the K'OTO project during the current COVID-19 pandemic.

1 Introduction

Triple helix players form the Mexican aerospace sector: Government, academy, and industry have made important efforts to increase existing capacities and generate conditions that allow the development of this industry at national and regional levels. Three specialized regions have been established in the country (center, northeast, and northwest) which leaves Mexico on the world stage, as a viable regional cluster of the aerospace sector. This includes infrastructure and existing services, specialized in human resources areas such as manufacturing, repair, and major maintenance, as well as engineering and design. Moreover, the cost of Mexican labor is considerably lower than in other countries; and the proximity to

S. O. Pérez-Elizondo
e-mail: soletae1998@comunidad.unam.mx

B. G. Reyes-Tesillo
e-mail: b.reyes@comunidad.unam.mx

D. Gaviria-Arcila
Advanced Technology Unit, School of Engineering, UNAM, Juriquilla Querétaro, Mexico
e-mail: dafne.gaviria@comunidad.unam.mx

the North American market, made our country an attractive place to invest in the aerospace industry since 2003.

The K'OTO project consists of the development of an educational nanosatellite with the purpose of Earth observation by remote sensing scanning all over the Mexican territory, promoted by the Secretary for Sustainable Development of the state of Querétaro (SEDESU). It is an initiative directed by academics and university students of the National Laboratory of Space and Automotive Engineering (LN-INGEA), of the High Technology Unit of the Faculty of Engineering of the National Autonomous University de Mexico (UAT FI-UNAM), Juriquilla Campus, where a team of students from diverse academic institutions, as well as from different areas and academic levels.

The expectations for great development for the space sector at a global level has generated new significant opportunities for developing countries. For this reason, a boost in the aerospace industry is forecasted in Mexico, therefore, the K'OTO project was planned to encourage the creation of proposal for the aerospace area in Mexico. The focus and priorities are the generation of human resources and the development of national technology, as well as in the awakening interest of the young multitude in this sector.

During the execution of this project, students from different universities such as UNAM (National Autonomous University of Mexico), UAQ (Autonomous University of Queretaro), ITQ (Technological Institute of Queretaro), ITD (Technological Institute of Durango), and UPSRJ (Polytechnic University of Santa Rosa Jáuregui) have been brought together, promoting the training of human resources in the space area through teamwork and multidisciplinary collaboration, providing practical knowledge to the students.

A common factor in Latin American projects is the lack of a public information strategy on the development, design, and manufacture, thus leaving limited documentation of their framework. Consequently, the interest of the members is to leave the evidence for future work, considering the management and engineering challenges of space systems applicable to the development of the K'OTO project, taking NASA's NPR 7120.8 as a reference.

The K'OTO project management involves workgroups that constitute different subsystems contemplated in the project, such as telecommunications, on-board computer, orientation control, telecommunications, Earth station, and structure, among others. By integrating the subsystems, the management of this project demands a comprehensive approach that allows for the possibility of tie actions between different knowledge areas.

2 Methodology

Throughout this space race, a variety of space agencies have generated standards for developments in this sector, whose objective is to produce an organized method to do things in the right way. The most referenced are the European Cooperation for Space Standardization (ECSS) and NASA's Systems Engineering Handbook.

Which one to use depends on distinct factors, such as the region of development, purpose of the mission, reviewers, and sponsors.

As described in Space Science Library and NASA, (2016),[1] "systems engineering" encompasses the methodical and multidisciplinary approach to the design, implementation, technical management, operations, and retirement of a system. The system is a set of hardware, software, equipment, facilities, personnel, processes, and procedures to achieve viable results for the performance and function of this structure in an interconnected way.

The methodology that will be followed for the development of this project, integrates practices for systems engineering and project management proposed by NASA, which incorporates models for the development, validation, delivery, and adaptation of updates to procedural requirements.

Complying with the standards and guidelines required for an evaluation of the project as it is through the Technology Readiness Level (TRL), proposed by NASA (Fig. 1) and stated in the NPR 7120.8 for the management of programs and projects that start from life cycles, Key Decision Point (KDP), and products evolving.

The project that is governed under this methodology has a Formulation Module that takes considerations of factors such as feasibility, risks, technology, and concepts; the approval from project stakeholders, those responsible for validating key point decisions; and the implementation of the approved plans that meets the approved requirements to proceed and evaluate them according to their planning and execution.

The deliverables that are in this methodology represent the minimum set necessary to carry out good management of programs and projects. The general

Technology Readiness Level (TRL) Definitions	
TRL 1	Basic principles observed and reported
TRL 2	Technology concept and/or application formulated
TRL 3	Analytical and experimental critical function and/or characteristic proof of concept
TRL 4	Component and/or breadboard validation in laboratory environment
TRL 5	Component and/or breadboard validation in relevant environment
TRL 6	System/subsystem model or prototype demonstration in a relevant environment
TRL 7	System prototype demonstration in an operational environment
TRL 8	Actual system completed and "flight qualified" through test and demonstration
TRL 9	Actual system flight proven through successful mission operations

Fig. 1 Technology readiness level definitions[2]

[1] NASA, NASA Systems Engineering Handbook, 2017, Nasa/sp-2016-6105 Rev2-Full Color Version, 12th Media Services, https://www.nasa.gov/sites/default/files/atoms/files/nasa_systems_engineering_handbook_0.pdf.

[2] NASA, (2013), NASA Strategic Space Technology Investment Plan, https://www.nasa.gov/sites/default/files/atoms/files/strategic_space_technology_investment_plan_508.pdf.

objective adopted for this project was the NPR 7120.8 suggested by the technical advisors. This is to simplify the life cycle and minimize the complexity of requirements for the type of applied research where, prototypes can be designed and tested or a simulation or model can be developed to demonstrate the potential of the research, leaving enough flexibility for innovation and creativity at the same time.

According to Eduardo, A. and Aguado, F.,[3] the life cycle for engineering systems is complemented by manuals from various international organizations, such as the International Council for Systems Engineering (INCOSE), the International Organization for Standardization (ISO), the International Electrotechnical Commission (IEC), and the Institute of Electrical and Electronics Engineers under standards such as ISO/IEC/IEEE 24748.

It is important to mention the limits of systems engineering ranging from technical, design, and configuration requirements, to system integration, analysis, verification, and validation (Fig. 2).

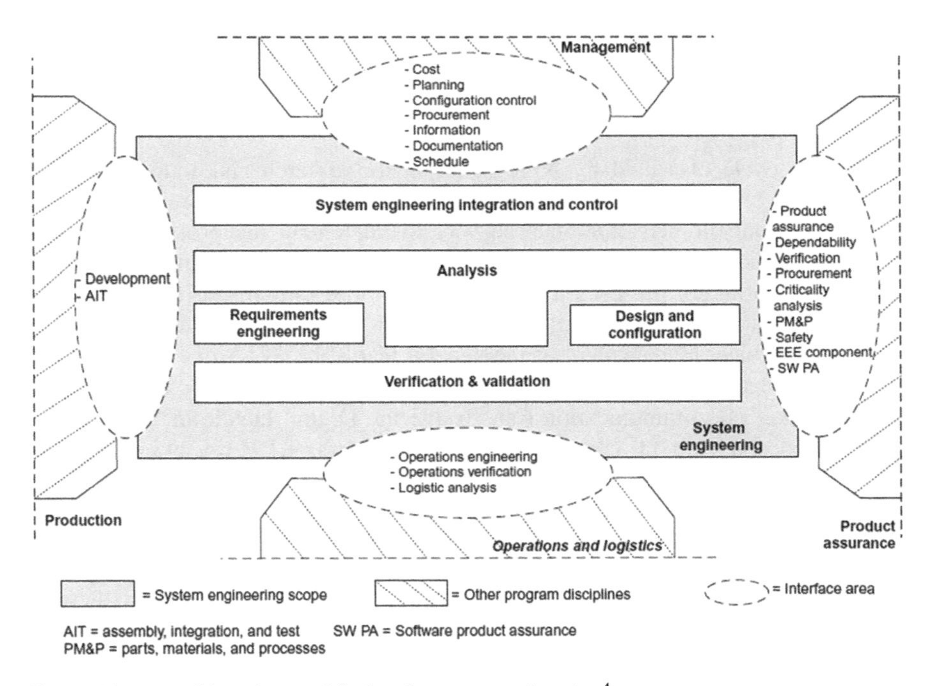

Fig. 2 Diagram of functions and limits of systems engineering[4]

---

[3] Eduardo, A. and Aguado, F., CubeSat Handbook, Part One: Systems engineering applied to CubeSats, 2020, 1st ed., Academic Press, https://doi.org/10.1016/B978-0-12-817884-3.09989-6. https://www.sciencedirect.com/book/9780128178843/cubesat-handbook#book-info.

[4] See footnote 3.

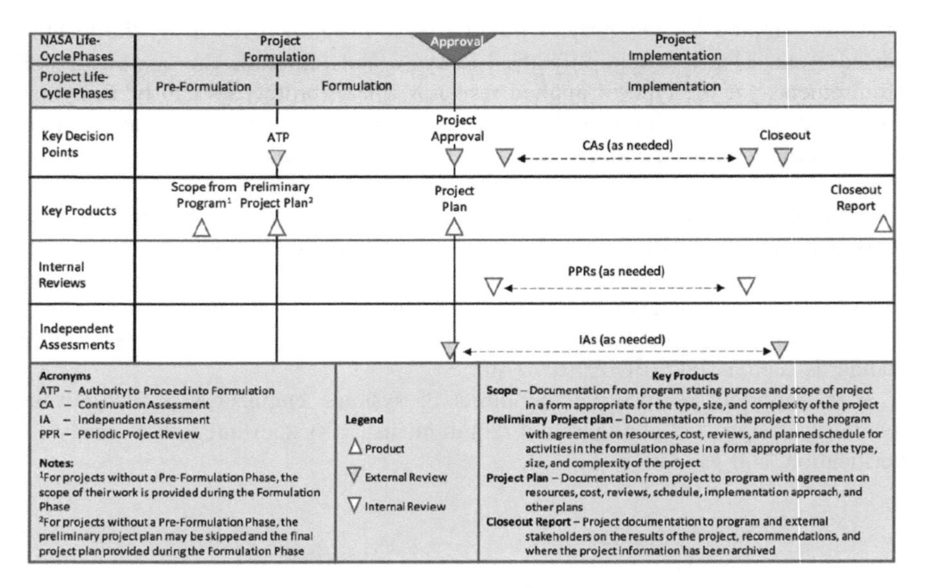

Fig. 3 R&T project life cycle according NASA (2018)[5]

The life cycle classified by NPR 7120.8A for research and/or technological development projects (R&T) is divided into three main phases: pre-formulation, formulation, and implementation. The way to implement the projects will vary between individual projects according to their type, size, and complexity. Each phase of the program life cycle includes one or more formalized reviews, whose objective are to evaluate periodically the technical and programmatic status to support the project management of the general plan (Fig. 3).

As described in NPR 7120.8A, chapter "Democracy Through Connectivity: How Satellite Telecommunication Can Bridge the Digital Divide in Latin America", to proceed with progress between these phases, KDP reviews should be considered which are flags that indicate the pass for the next phase of the life cycle. This includes Authority to Proceed (ATP), Program Approval, Program Evaluation Reviews (PAR), and the closure of the project.

When starting any space project, specifically for a CubeSat mission, it is essential to define in a quantifiable and verifiable way the needs or requirements that are expected to satisfy the objectives, set for the design, integration, and operation of the satellite.

The scope will indicate the complexity and challenges of the project considering the program from the sponsors, technical programming, and the work team, as well as technology transfer, costs, quality, and schedule. This will allow that even at a

[5] NASA (2018), R&T Project Life Cycle, [Fig. 3–1], in NASA Research and Technology Program and Project Management Requirements, NPR 7120.8A (2nd ed., p. 15).

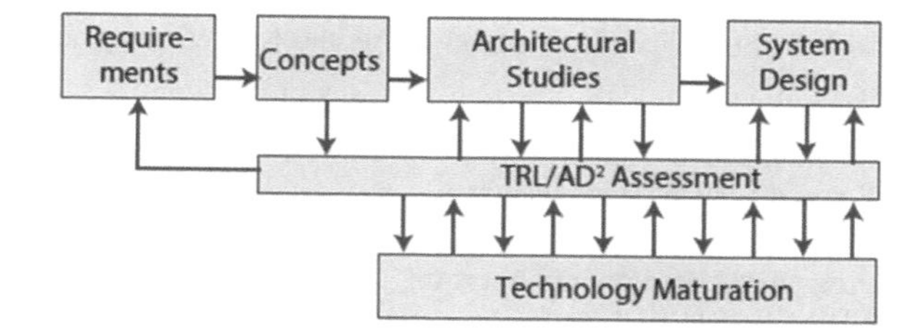

Fig. 4 Technology assessment process for architectural studies and technology development[6]

conceptual level, architecture is designed for the Technology Assessment Process (Fig. 4), which will be intertwined to avoid errors as much as possible.

Thus, it is possible to refine the system design for optimal performance in the meeting objectives of the proposed mission. The constant documentation of the subsystems is recommended, where the progress in the level of maturity reached during each review is observed, as well as the planning of alternative paths and the identification of new areas are required for custom development to translate into the following deliverables.

Within the Flight Design Specification of this satellite, the published standards must be considered according to the program that will be adopted at the beginning of the project to go within the NORM, as mentioned in chapter "Social Sustainability: A Challenge for the Supply Chain of the Mexican Space Sector" of the Requirements definition: User, mission, and system by Eduardo and Aguado (2020).[7] Some usual standards are the JAXA JEM Payload Accommodation Manual, the ESA Space Debris Mitigation Requirements, ECSS-U-AS-10C, the Regulation of ITU Radio communications, the NASA's Safety Requirements for Manned Space Vehicle Batteries, the ISO 17770-Space Systems Hub Satellite Standards.

For the precise and accurate development of the project, a group of experienced authorities will be assigned to help and manage the mission to have good decisions through reviews and evaluations carried out throughout. This group of people is designated as the Partner Center Board of Directors, described in the Formulation Phase. The following Table 1 lists some of the authority roles to consider within the project to be developed[8]. It should be noted that the number of these directors varies according to the complexity of the project.

The next deliverable document corresponds to the Preliminary Design Review (PDR) whose purpose is to evaluate and validate the design that is following the requirements set out at the beginning, the acceptable risk, and within the limitations of cost and schedule. During the development of the mission, a periodic

[6] Adapted from Space Science Library and NASA, (2016), NASA Systems Engineering Handbook - NASA SP-2016–6105 Rev 2: Design Test Integrate Fly, https://www.nasa.gov/sites/default/files/atoms/files/nasa_systems_engineering_handbook_0.pdf.

[7] Eduardo, A. and Aguado, F., CubeSat Handbook, Part One: Systems engineering applied to CubeSats, 2020, 1st ed., pp. 11–15, https://doi.org/10.1016/B978-0-12-817884-3.09989-6. https://www.sciencedirect.com/book/9780128178843/cubesat-handbook#book-info.

Table 1 Summary of governance authorities for R&T programs[8]

Authority Role	Authority	Comments
Program DA for Authority to Proceed KDP and Formulation Phase	MDAA	
Program DA for Program Approval KDP, Closeout KDP, and Implementation Phase	NASA AA	The NASA AA can delegate responsibility to the MDAA. The DA may request additional KDPs (PARs) during Implementation
Management Official for Establishing Independent Assessment Team(s)	MDAA	The MDAA identifies the chair of the independent assessment team. The chair selects any team members. Approvals/concurrences are obtained from the Mission Committee, implementing Centers, and OCE for the (1) IA Chair, (2) IA team members, and (3) IA expectations document. The MDAA will ensure that the team(s) and process are independent and objective
Governing PMC for Formulation	DPMC	
Governing PMC for Implementation	APMC	The NASA AA can delegate oversight responsibility to the DPMC
Governing Document	R&T Program Plan	The R&T Program Plan is approved by the MDAA

review of the project and the validation of the reports by each subsystem will be conducted (Fig. 5). Finally, the plan describes the relationship between the communication plans of the schedule and the progress of the project.[9]

The Project Plan contains a series of mandatory control plans with the ones that are required during this phase, the Control Plans are required to be developed, taking as reference the latest version of the PDR should be described. These early control plans streamline work management and verify budgets, time, quality, and expected results, considering the KDP's, being for Constituent Parts of the Life Cycle Cost Estimation of a Project for Formulation and Implementation.

The Project Plan control method found in the baseline governs the management and technical control processes used during this phase. Based on the refined

[8] Adapted from NASA, (2018), Summary of Governance Authorities for R&T Programs, NASA Research and Technology Program and Project Management Requirements, https://nodis3.gsfc.nasa.gov/displayDir.cfm?Internal_ID=N_PR_7120_008A_&page_name=Chapter3.

[9] National Aeronautics Space Administration (NASA), General Environmental Verification Standard (GEVS) for GSFC Flight Programs and Projects, 2021, https://www.cubesat.org/cubesatinfo.

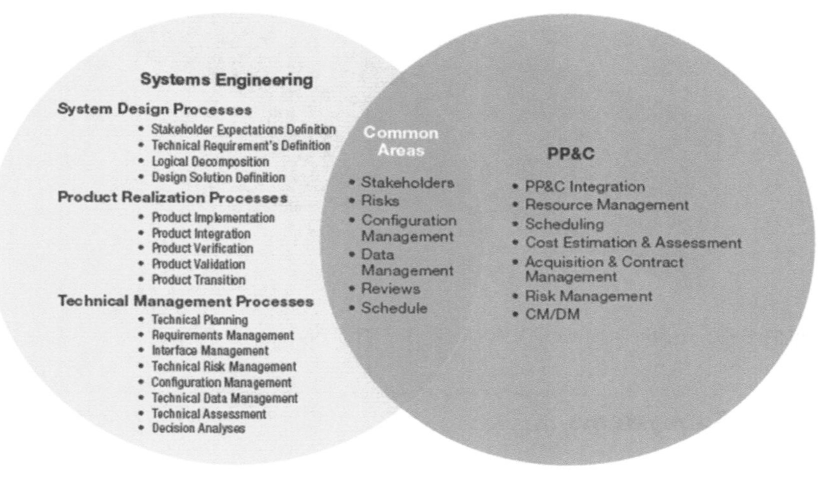

Fig. 5 System engineering in context of overall project management[10]

concept/design and its exposure, a risk-informed schedule is developed at the system level (at a minimum) with a preliminary estimate of the completion date and analysis of new requirements. This is in conjunction with the budget and with this visualization of threats in the development of the system with their respective mitigation plans.

The Project Plan is constantly maintained and updated, until reaching the closing, where the elimination or destination of the components that made up the project is considered, such as Earth station equipment, test benches, and spacecraft, among others.

When these reports are generated, it informs about open and possible risks mitigations, complete mission file, operational-scientific data, technical and administrative documentation. Further on the respective evaluation of mission and removal activities, program achievements, summary of program metrics, rationale for program closure, lessons learned, archiving, storage, deletion, security approach to program information and artifacts and technologies are developed.

[10] See footnote 6.

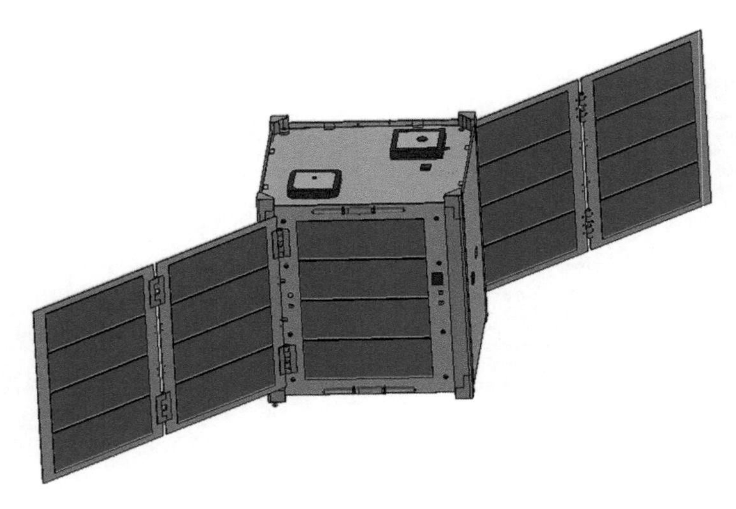

Fig. 6 Basic satellite structure CubeSat 1U (K'OTO)

3 Subsystems

3.1 Structure Simulation Analysis of a CubeSat

The structure of a satellite is one of the important tasks that a team who develops a satellite must achieve in all the cycle life of a satellite. The principal objective of the structure is to assure the arrangement of all components inside as well as to guarantee the survival of all the systems inside the panels due to the space radiation and rocket vibration environment when a satellite is on board. One example of a satellite structure CubeSat 1U is shown in Fig. 6.

This chapter describes just how to perform a simulation analysis by Finite Element Analysis (FEA). The type of analysis that is described here are static analysis, modal analysis, and random vibration. The development of this work is with the license of the software ANSYS.[11]

3.2 Simulation Analysis

All procedures here for the simulation analysis follow the requirements from JAXA for small satellites. As is not any rule method for doing FEA analysis for satellite structures, you can find spread documentation about this topic, but every development of a satellite is a unique case. The best practices are always to have a fundamental theory about the task analysis for each structure's requirements. For

[11] Sponsored by ANSYS Grupo SSC.

this chapter, the most important theory topics to success in a simulation with these conditions described before are the following:

- Mechanical vibration.[12]
- Fundamental of Finite Element Analysis.[13]
- Basic concepts of ANSYS software.
- JAXA procedures manuals and standards.[14]
- NASA procedures and manuals standards.
- CubeSat standards.

3.3 Geometry Preparation the First Steps for Simulation

Geometry preparation is a key step to do before doing simulation analysis. If the clean geometry is clean the equation will be solved faster, importing models from any CAD design software is not always the best practice for systems that have many components. It becomes worst when the geometry comes from many reworks by different computing systems due to data loss of fracture information. ANSYS has the capabilities to add patches and fix some geometry problems automatically or even manually, but these are not the best practices when solving the equations models and is going to take a lot of time and computer resources to solve.

The best practice (if possible) is to draw again the model in a simplification way with a native software design modeler embedded inside the software that is going to solve the FEA analysis. ANSYS has a package that allows drawing in multiple dimensions (1D, 2D, and 3D), making it possible to use that model to solve the mathematical equation in a faster way.

If there is not an indigenous way to model into the software FEA tool it is possible to import the geometry data. Whether the engineer decides to redraw or import the geometry it is important to apply the next tips to become or draw a good geometry shape.

The first step before doing FEA analysis is always to avoid fasteners, screws, threaded holes, small or all radius, sharps, printed details, and components that are not representative of the model analysis. Whether the modeler prefers to draw or clean the geometry needs to keep this in mind.

Figure 7 represents a simplification geometry of Fig. 6. There are no details components, and some parts of the deployment panels are replaced for components that do not represent an interest for this analysis but play a key role in

[12] Wijker, J.J., Mechanical vibrations in spacecraft design, 2013, Springer Science and Business Media.

[13] Abdelal, G.F., Abuelfoutouh, N., and Gad, A. H., Finite element analysis for satellite structures: applications to their design, manufacture, and testing, Springer Science and Business Media, 2012.

[14] Handbook, J.P.A., Vol. 8 - Small Satellite Deployment Interface Control Document, 2015, Japan Aerospace Exploration Agency (JAXA).

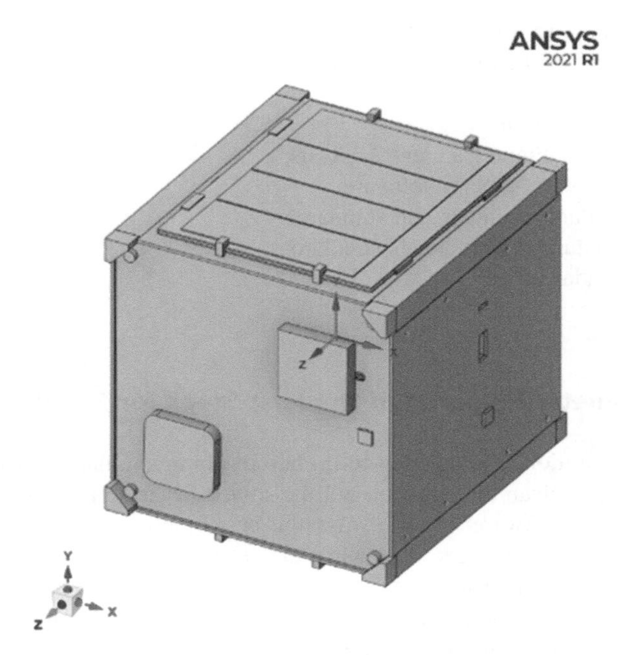

Fig. 7 K'OTO Satellite 3D model for FEA–outside view

giving stiffness to the system. Figure 8 represents the internal component in a simplification shape, the model has no wiring or components details, and some parts have different geometry that cannot affect the results.

Creativity is not limited here, there are many ways to represent and simulate FEA systems. Each case is going to depend on the engineer modeler and the kind of analysis to be solved.

3.4 Material Properties and Mass Distribution

Before solving any simulation, it is necessary to set the materials parameters and mass distribution. As this analysis is just focused on the analysis of the structure, the electronic and internal components can be suppressed from the model. To have many components that are not representative of simulation are going to consume a lot of computational resources and time of solution. Because mass and mass distribution are important for a vibration analysis some technics can help to replace the model CAD components and will not affect the results. Several tools can allow for mass replacement components and fasteners screws.

Virtual mass is a valuable tool to replace mass components, these tools add mass points in any part of the geometry and set a support base area. As much detail the engineering designer wants to have the more point masses can add to

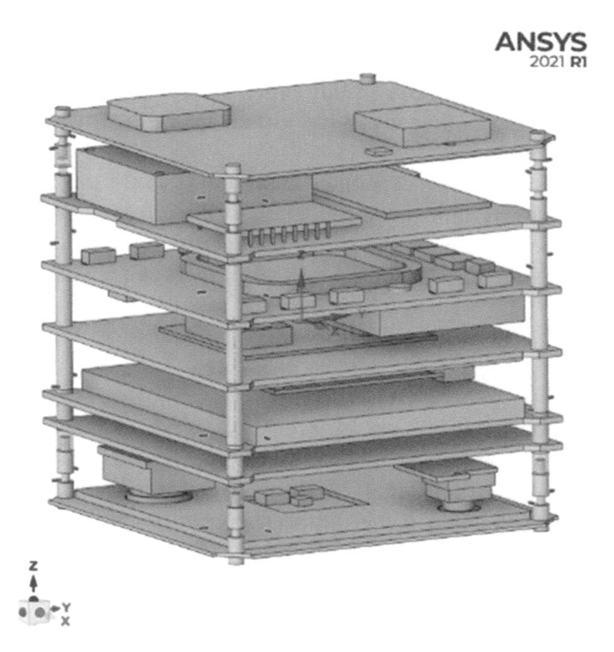

Fig. 8 K'OTO Satellite 3D model for FEA–inside view

the model. The important factor here is to keep the global center of mass gravity. Figure 9 represents an example of mass distribution.

The example in Fig. 9 is not a mandatory way to represent an FEA system. If the engineer requires more details in any specific component, it is free to be added to the model.

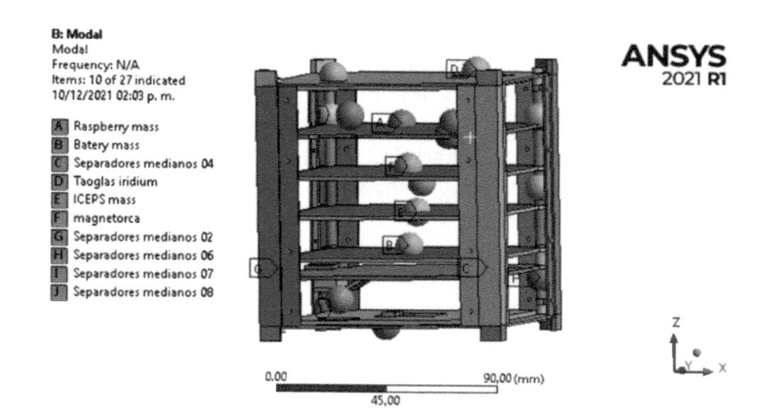

Fig. 9 Virtual mass distribution of model K'OTO

The principle of Saint–Venant[15] is one of the most key factors in FEA analysis. This principle gives us the freedom to solve with special detail in any region of interest without refining mesh in the entire system. Table 2 shows the global values of the mass distribution involving virtual mass plus mass materials of the components.

Adding materials properties is another important task, for the equations that represent the mathematic systems need to take the value of young modulus, strength, and Poisson relation. Figure 10 shows an example of some materials you can use in the model and Table 3 gives some material properties for simulation analysis.

3.5 Mesh Quality

For CubeSat geometries the most representative mesh elements are hexahedral. When there is more complex geometry, like for example some parts of the riel the model has tetrahedral elements. Mesh quality is a tool that allows measuring the quality of every element in the mesh. There are many ways to evaluate and measure the quality of the elements. Choosing the best parameter measure will depend on the model shape, it can be modeled with shell surfaces, plates, beams, etc.

An example of a mesh parameter that can't fit well for plates parts is the *element quality*. This parameter measures how an element is close to being a perfect shape like squares, cubes, triangles, tetrahedral, etc. Sometimes if the engineer wants to have hexahedral shapes elements inside a plate part with few elements despite a good and clean geometry shape, is not possible to get perfect cubes because the relation between length, width, and depth of the plate is very big and probably instead of having perfect cubes there are perfect rectangles shapes, which are not a standard parameter for *element quality* tool. Then probably this parameter is not the best for plate shapes as is shown in Fig. 11.

As closer to the maximum of 1,00 in the color scale in Fig. 11, the more perfect the geometry shape of the elements are. This example shows that *element quality* is not the best way to measure the mesh quality.

Another parameter that works better for this shape is the *parallel deviation*. This parameter tells if the opposite edges of elements are as parallel as possible. This parameter works only for hexahedral elements. Figure 12 shows how this parameter admits cubes and rectangle shapes. As much close to the zero in the number scale the more parallel edges the mesh elements are.

[15] Dou, A., On the principle of Saint-Venant, Wisconsin Univ. Madison Mathematics Research Center, 1964.

Table 2 General Mass, Centroid, and Inertial Moment of model K'OTO

Mass	Centroid			Inertial moment					
[g]	X [mm]	Y [mm]	Z [mm]	I_{xx} [Kg mm^2]	I_{yy} [Kg mm^2]	I_{zz} [Kg mm^2]	I_{xy} [Kg mm^2]	I_{yz} [Kg mm^2]	I_{zx} [Kg mm^2]
1.030,9	−1,048	−0,906	−3,081	2.682,4	2.292,5	2.506,5	−9,432	47,748	−21,253

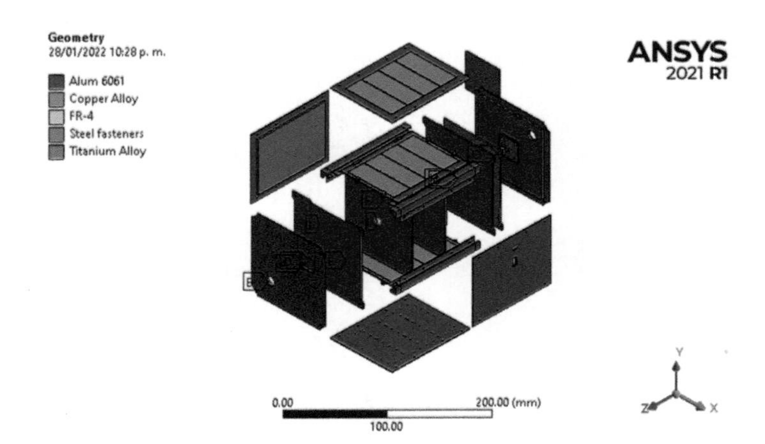

Fig. 10 Representation of material properties in the model K'OTO

Table 3 Material properties of model K'OTO

Color	Material	Young modulus [GPa]	Poisson relation	Tensile strength [MPa]	Ultimate strength [MPa]	Density [Kg/m^3]
Blue	Aluminum 6061	68,9	0,33	276	310	2,700
Red	Titanium	96	0,36	930	1.070	4,620
Yellow	FR-4	23,4	0,14	275,79	310	1,800
Red	Cupper	110	0,34	280	430	8,300
Gray	Steel A-286	206,8	0,31	827	1.100	8,310

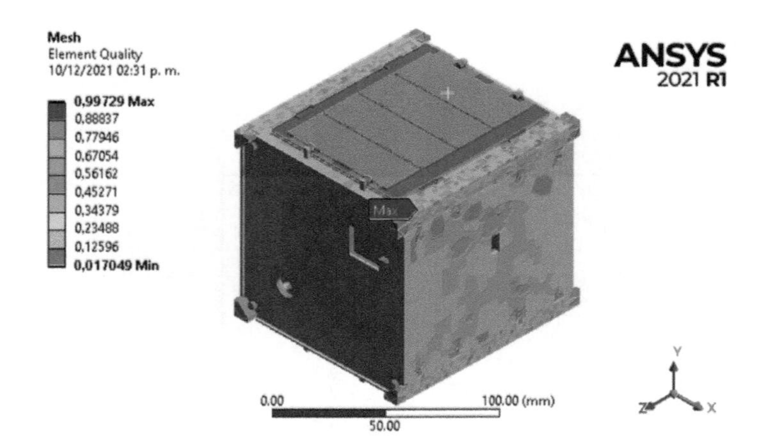

Fig. 11 Element quality graphic representation of model K'OTO

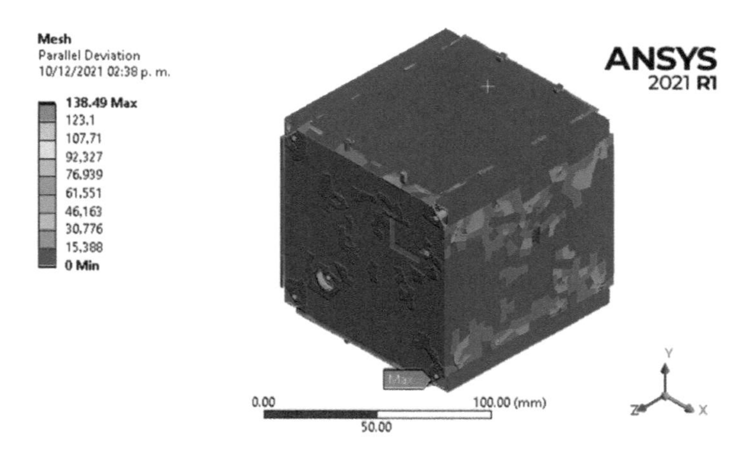

Fig. 12 Parallel deviation graphic representation of model K'OTO

3.6 Independence Mesh and Convergence Solutions Technics

This chapter is going to take an example of the development of the K'OTO satellite as a satellite with the purpose to develop new professional people in the space sector.

As the beginning of this chapter said, there are many methods to find the independence mesh and a convergence solution stress number. This next literature will try to explain some steps and recommendations to follow to have good analysis results as well as the engineer is trying to find the maximum stress value in every iteration. These can work for many types of analysis but in this case, is just about static and vibration analysis. The forces applied are shown in Fig. 13.

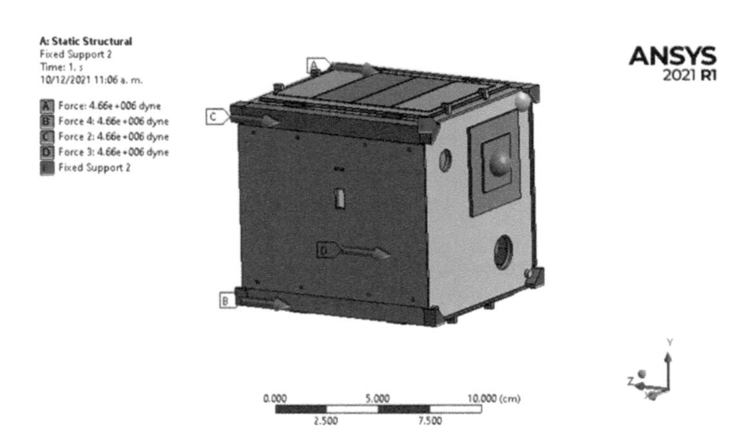

Fig. 13 Forces and fixed conditions for static simulation analysis of model K'OTO

After that, the model is cleaned, and it has decent quality mesh with a property parameter measure tool, only at that moment, the contacts can be added. Bonded contact uses linear equations and is a useful tool to start solving the model fast. Bonded contact restricts the six degrees of freedom of the contact surface. Another important fact to be considered for this satellite example is that, if you observe the assembled structure and how forces are applied there, is going to be a displacement between the external plate and the riel part, with bonded contact is not possible to see this reaction. To solve this problem, it is better to use no separation contact. This contact also uses linear equations and allows the movement without any restriction in a tangent direction between the two surfaces in contact.

It is necessary to do some iterations starting with a coarse mesh and refining the mesh size in every iteration. For now, is not important to find the value of the maximum stress, the most important task is to find the zone where the maximum stress is. If the maximum stress change in every step, go back to the previous steps and try to improve either the geometry, border parameters, mesh quality, etc.

After the zone of maximum stress is found, the next step is to find the value of stress. In FEA analysis there is not an exact value to find as in an analytical method. It is going to exist always an error in every iteration solution found when the mesh is refined, the thing is to reduce that error to under 3% or 5% in each iteration.[16]

The cycle of the design follows some rules before delivering the fly model structure. Figure 14 shows a diagram with the steps of a cycle life of a product for systems engineering but it can be found in any mechanical design book literature. The diagram shows that if the simulation results do not meet good requirements is necessary to go back in the design model and change geometry shapes in the critical area that has a high risk of failure. If the simulation requirements meet reliable results, it can meet a prototype, this can be also called the engineering model. This model has to be tested to validate the FEA simulations analysis.

In space engineering testing a prototype or engineering model is a very important step that allows giving reliability to the structure system.

When the testing results match with simulation results sometimes is needed to make new simulations for random vibration, this is because is not possible to know the real amplification factor Q of the structure by simulation. After the amplification factor is calculated from the testing analysis, it needs to be updated and set for new simulation analysis. Finally, when simulation and testing satisfy reliable results the fly model can be manufactured and ready for launch and put into orbit in space.

Going back to find the maximum stress value and following the technics as was mentioned previously, the next step is to change the contact parameter from bonded and no separation to frictional contact only in the zone of interest where the maximum stress value is found. It is important to change all the contacts that

[16] American Institute of Aeronautics and Astronautics, Guide for the verification and validation of computational fluid dynamics simulation, AIAA, 1998.

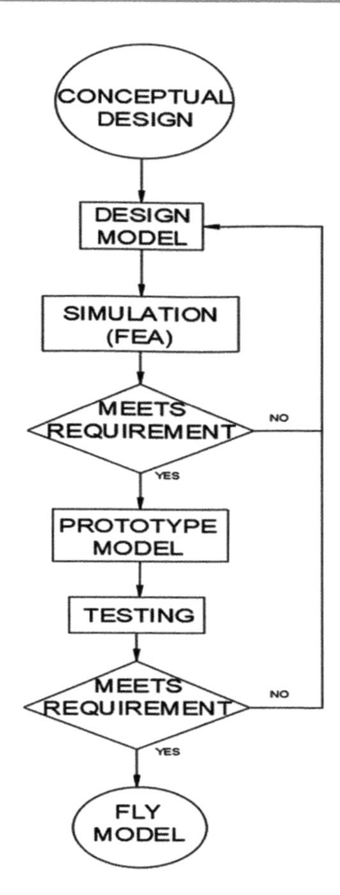

Fig. 14 Design cycle life of a generic product for systems engineering

interact with the zone of interest, Fig. 15 shows the zone with the maximum stress and the area that needs to be worked on.

One typical problem that can be found is to have stress singularities in the model, this is because the model has perfect rectangle shapes and singularities occur in corners or edges. Is easy to confuse stress concentration with stress singularity because both can increase the stress in high order.

If in every iteration of refining mesh, the values of stress and number of nodes are plotted, it is possible to see how stress concentration rich a convergence value, but stress singularity will never reach a convergence value, and is going to increase infinitely as much smaller the size of the elements are.

A way to solve the problem of stress singularity is by adding the real radius to the model just in the zone of interest and stress singularity is going to disappear. But another technic is to use a path that crosses the zone of interest (zone of maximum stress) as shown in Fig. 16.

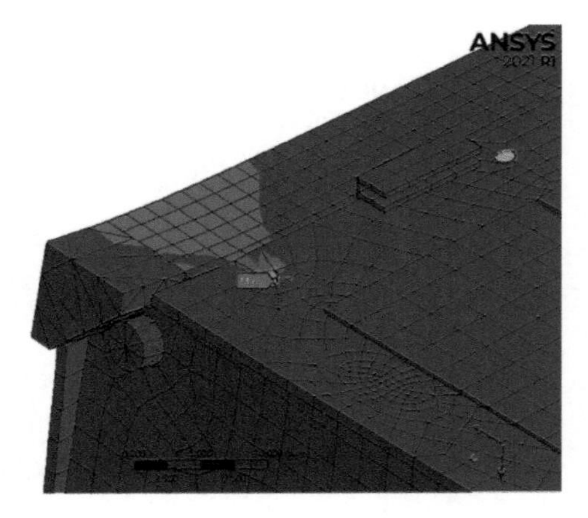

Fig. 15 Zone with maximum stress value of model K'OTO

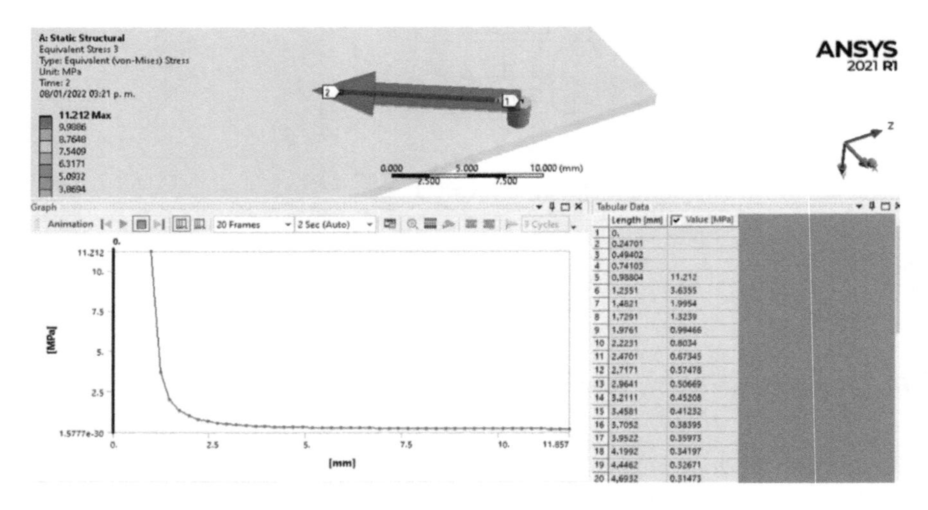

	Length [mm]	Value [MPa]
1	0.	
2	0.24701	
3	0.49402	
4	0.74103	
5	0.98804	11.212
6	1.2351	3.6355
7	1.4821	1.9954
8	1.7291	1.3239
9	1.9761	0.99466
10	2.2231	0.8034
11	2.4701	0.67345
12	2.7171	0.57478
13	2.9641	0.50669
14	3.2111	0.45208
15	3.4581	0.41232
16	3.7052	0.38395
17	3.9522	0.35973
18	4.1992	0.34197
19	4.4462	0.32671
20	4.6932	0.31473

Fig. 16 Path method of model K'OTO

According to the principle of Saint-Venant[17] is possible just to avoid singularities, and we know that in real-life perfect corners do not exist because there are no machines that allow us to have those perfect parameters. The path method shows the stress at every point in the path, in that way we can avoid the points that are inside singularities and take the point that must settle down values in each iteration. Finally, refining just the local area is going to save computational resources and give good solutions.

[17] See footnote 11.

3.7 Modal Vibration and Random Analysis

Modal vibration is a linear analysis, then the use of only linear contacts is allowed for this method. The important thing here is to set the mass distribution and stiffness parameters well. The virtual mass in Fig. 9 can be used to find the modal vibration.

The participation factor and the ratio of effective mass vs total mass of a modal analysis can help to find the natural frequency and the minimum vibration modes. Figure 17 shows an example of how modal frequency results look and Fig. 18 shows an example of the ratio of effective mass vs total mass of the K'OTO satellite.

Finally, after the minimum modal frequency is found, the random vibration will be the next step in this analysis. For random vibration, the Power Spectral Density (PSD) needs to be set it down as a new parameter, one example is shown in Fig. 19.

It is possible to follow and repeat the same formula as mentioned in the previous chapters:

1. To use coarse mesh in the first iteration.
2. To use linear contact.
3. To find the maximum stress zone in the first iteration.
4. To work only in local zones and uses local refine mesh area.
5. Avoid stress singularity.
6. Graph stress vs number of nodes elements.

For the K'OTO example, the zone of interest is shown in Fig. 20.

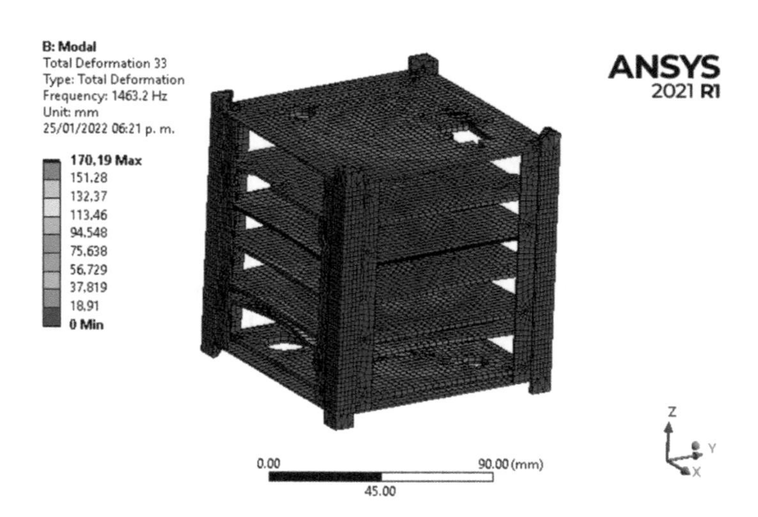

Fig. 17 Modal frequency of model K'OTO

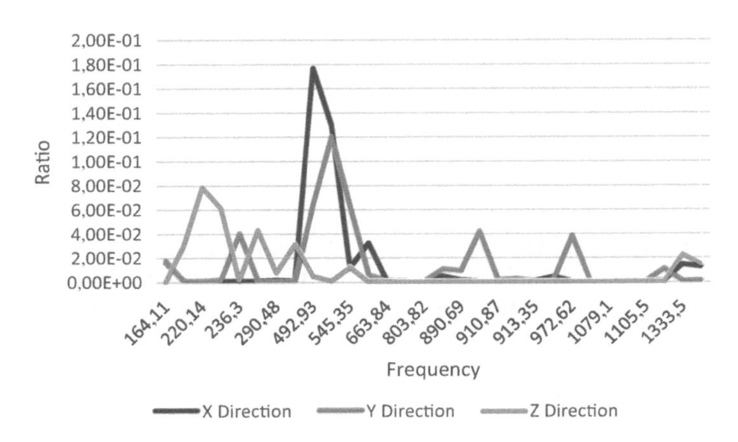

Fig. 18 Ratio of effective mass versus total mass of model K'OTO

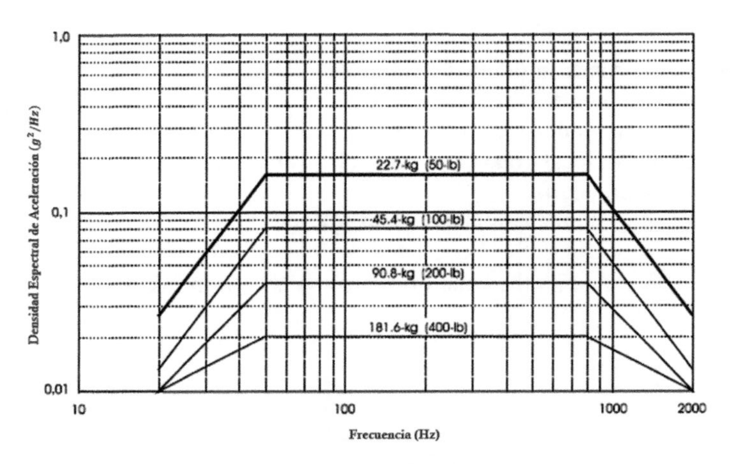

Fig. 19 Graphic of the typical power spectral density of random vibration described in GSFC-STD-7000A[18] and used in the model of K'OTO

Finally, it is important to find and graph the PSD response in one external surface point of the structure. This graph can help the validation of the PSD response between the simulation analysis and testing values. Figure 21 shows an example how the PSD response looks like at a point located in +X external face.

[18] NASA Goddard Space Flight Center, Greenbelt, Maryland 20771 (2019), General Environmental Verification Standards (GEVS) GSFC-STD-7000A.

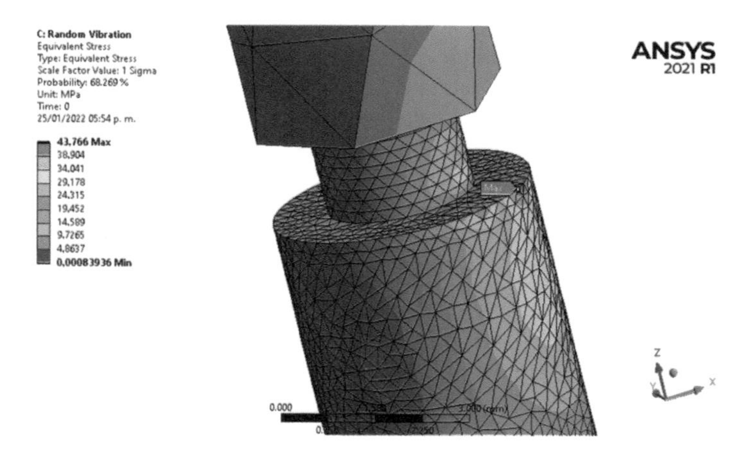

Fig. 20 Zone of interest of model K'OTO

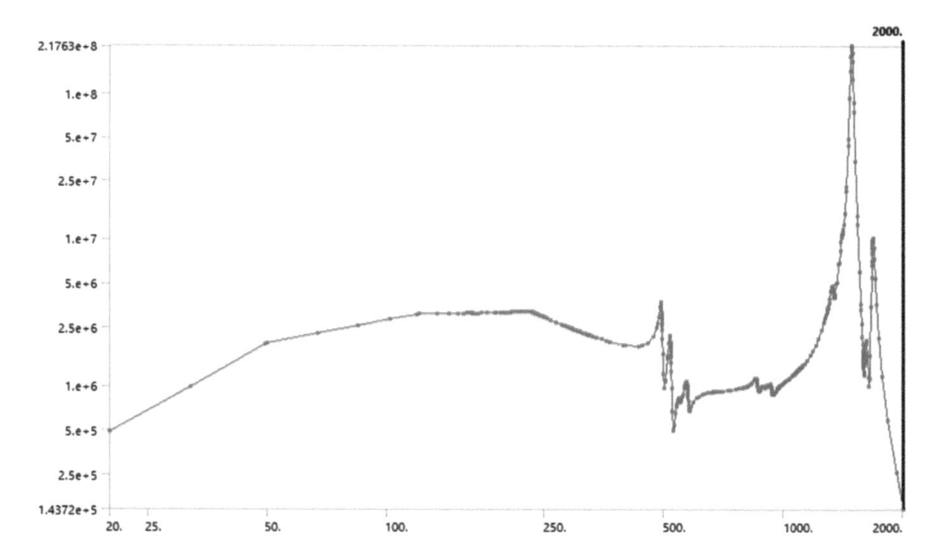

Fig. 21 PSD Response in +X face of model K'OTO

4 On-Board Computer

The On-Board Computer (OBC) is the subsystem that must integrate the satellite subsystem modes to fulfill the mission requirements. The operations on the OBC are defined along the interface design with other subsystems. The design is then constrained by such operations, budget, and flight modes. Therefore, the next considerations must be analyzed to design a suitable solution for the OBC design.

One of the main operations to be executed in the space element is Command and Data Handling (C&DH), including command packing, unpacking, validation, and interpretation. The operation is partially resolved between the communication and OBC subsystems where the commands then trigger a change in the satellite mode. Even more, the OBC is also the intermediate system between the instructions and all the applications in the space segment.

Also, Data Handling involves the recollection and transmission of the housekeeping data and telemetry.[19] One option to solve this is by reading data from sensors linked directly to the OBC. The link can be either analog or digital with another component attached to the OBC design. Another option is to store the system-related data in a subsystem or module itself, and later integrate the information with the OBC at the required moment. These strategies are not mutually exclusive and can have different data gathering procedures. This is because data recollection algorithms or technologies might have different time delays or data availability time lapses.

The total amount of tasks that the OBC must accomplish is what settles the requirements for the performance level of the subsystem. Such requirements include memory size, processing speed, architecture, electronic interfaces, robustness, and compatibility with the rest of the spacecraft.[20]

Other important considerations are the structure, energy consumption, thermal characteristics, and electromagnetic compatibility. Different missions might require several types of technologies interfacing with the OBC. Therefore, is important that the type of technology selected for the OBC is compatible with the rest.

All these considerations show that the OBC design and implementation are not isolated. Rather, those demand several iterations to adapt all the modes and any changes in the other subsystems. The different interactions and restrictions defined by each subsystem are specific to each project. Everything must be detailed in the documentation as requirements. In the beginning, only general guidelines are stated, then more detailed definitions of the operation modes, hardware configuration, standardization of data, and commands specifications.

The most general agreement is the mode specification for the spacecraft. Also, each mode can be broken down into subsystem modes. All of this is written down in a timetable where the trigger events and mode changes can be mapped together. In this representation, the rows match a subsystem, and the columns separate the consecutive modes in a timeline. The diagram can be as detailed as needed. However, do not forget that in each phase of the NASA life cycle the diagram has to be updated (Fig. 22).

[19] Wertz R. J., Everett F. D. and Puschell J. J., Space Mission Engineering: The New SMAD, Hawthorne, CA, Microcosm Press, 2011.

[20] Eickhoff, J., Onboard computers, onboard software, and satellite operations: an introduction, Springer Science and Business Media, 2011.

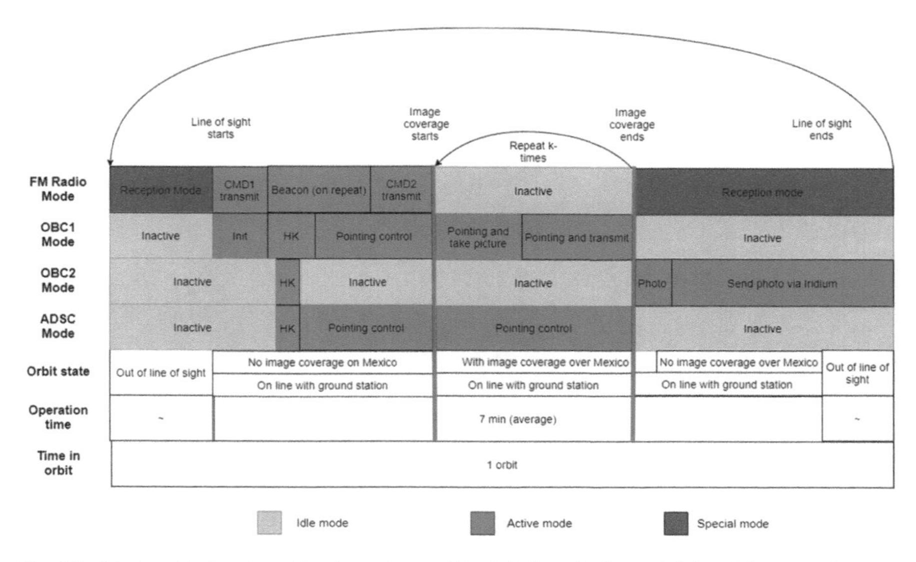

Fig. 22 Mode table for the orbit where the satellite falls into the line of sight of the ground station

It is encouraged to use the different Unified Model Language (UML) diagrams[21] available to portray information and agreements.

The documentation should have information regarding the subsystem interactions within each other. This is established in the UML interface diagram. In this diagram, the main guides between subsystems are defined by the master devices in each interaction. Notice that communication buses are not specified. This diagram helps in the concept of the functionality of the software, but different buses might be implemented to fulfill the requirements in this diagram. In other words, this is a framework for the application development in each component on the satellite (Fig. 23).

This Framework has shown that Commercial Off-The-Shelf (COTS) electronics are suitable for short missions in a Low Earth Orbit (LEO).[22] This allows the use of COTS components for electronic design. Therefore, LEO is useful for educational projects and testing innovative technology. In other words, there is a large variety of options for a processing unit for educational purposes spacecraft that require a lower investment than other types of satellites.

Even more, there are different technologies available for the processing units. Educational nanosatellites have a wide range of performance needs. Simple tasks

[21] Kim Hamilton, R.M., Learning UML 2.0, Publisher: O'Reilly, 2006.

[22] de Oliveira, J.C. and Manea, S., Methodology for the Selection of Cots Components in Small Satellite Projects, and Short-Term Missions.

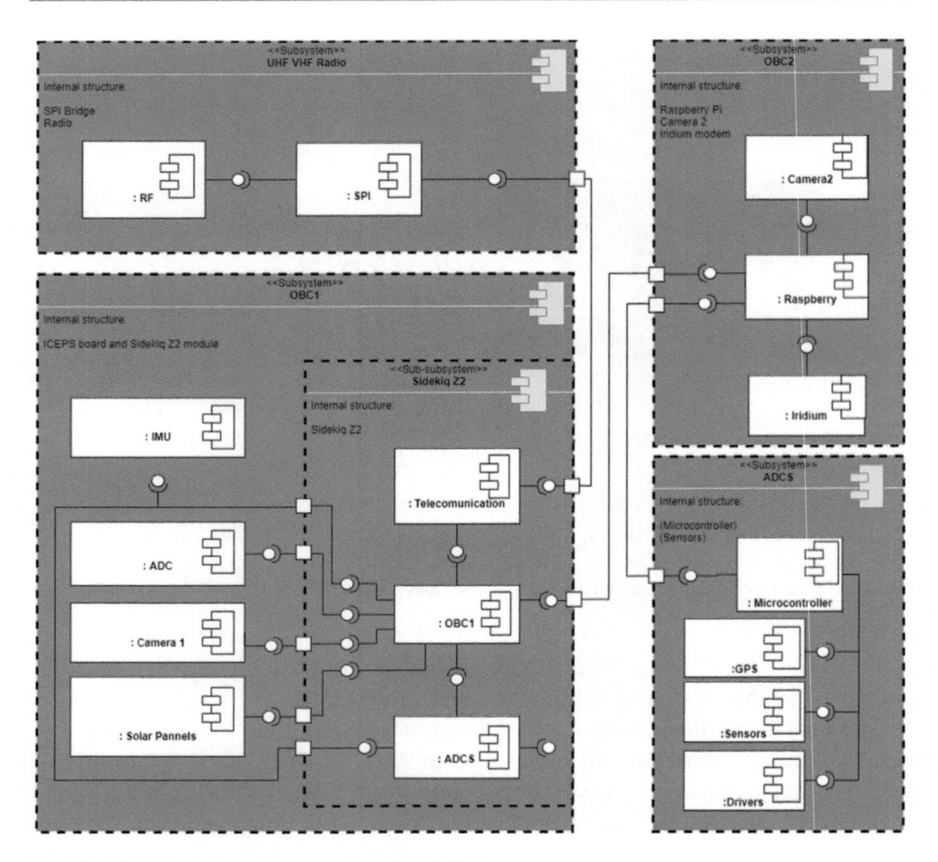

Fig. 23 UML interface diagram of model K'OTO

have been executed by microchip PIC microcontrollers in the BIRDS BUS[23] and Atmel microcontrollers for projects like the Interorbital bus in Tancredo-1.[24] More complex tasks can be performed by microprocessors. Such is the case of the SPUTNIX bus that is based on a "Raspberry Pi" which uses an ARM microprocessor. This has proven to be suitable in the cases of SiriusSat-1 and 2.[25] Also, technologies like FPGAs are the base of the ICEPS BUS.[26]

[23] Kim, S., Yamauchi, T., Masui, H., and Cho, M., BIRDS BUS: A Standard CubeSat BUS for an Annual Educational Satellite Project, JoSS, 10, pp. 1015–1034, 2021.

[24] de Oliveira, J. C. and Manea, S., Methodology for the Selection of Cots Components in Small Satellite Projects, and Short-Term Missions.

[25] Bogomolov, V.V., Bogomolov, A.V., Dement'ev, Y.N., Eremeev, V.E., Zharkih, R.N., Iuydin, A.F., … and Svertilov, S.I., A scientific and educational space experiment on the Siriussat-1, 2 satellites, Moscow University Physics Bulletin, 75(6), pp. 665–675, 2020.

[26] Nader, R., Nader-Drouet, J, and Nader-Drouet, G., ICEPS: Compact, all-purpose, USB 2.0 based small satellite system core, 2019.

In the end, the options for suitable OBC hardware are diverse. Then, the final criteria are any other secondary objectives that can be set to the main mission. In the case of K'OTO, the use of the ICEPS kit give the hardware for the OBC subsystem. However, we decided to use a second OBC based on the "Raspberry Pi" as a redundant OBC. Each computer can perform the mission sequence, which consists of the orientation of the spacecraft, capturing images of the Mexican territory, and a later transmission of the photos to Earth.

One computer (OBC1) is based on the ICEPS bus that hosts a Sidekiq Z2 SDR with IIO Linux. The firmware runs on a Cortex A9 processor synthesized in an FPGA attached to the radio. This means that the processing unit and the SDR are in close communication with each other. Also, OBC1 operates one of the payload cameras. In other words, this computer is testing both this particular SDR in S-band and the Leopard camera in a space environment. There are two other tasks that this computer must accomplish in this case study. One task is to compute one of the ADCS algorithms, taking advantage of the higher processing capabilities. Another task for this computer is starting the deployment sequence right after the launch.

Conversely, the other computer (OBC2) is based on a "Raspberry Pi" zero with an ARM 11. It is equipped with a Raspberry camera v2 and a communication modem for backup communication. The modem considered is from enterprise Iridium®, which is a private telecommunication provider based on its constellation. The OBC2 can take pictures and send them to the ground station using the Iridium modem. OBC2 is the redundant subsystem of the K'OTO nanosatellite, designed to take control of the nanosatellite in case of a malfunction event of OBC1. The possible risks are malfunctions of the OBC1 processing unit, camera, and/or the communications subsystem. Because of this, OBC2 has its camera and communication module. Even more, all components in this subsystem have flight inheritance to minimize the probability of failure. In other words, the main purpose of this subsystem is to ensure that K'OTO can fulfill its mission even if unforeseen events occur. In addition, even when the main subsystems are working properly, OBC2 provides additional operability to the satellite. Commands sent from the ground station via the Iridium module can trigger the process of capturing and sending pictures.

The main difference between both computers is the complexity of the tasks. OBC2 is designed to use slimmer algorithms and operations to reduce the possibility of failure. In addition, OBC2 does not depend on the same camera or communication systems as OBC1.

Both OBC's engineering models have been implemented to conduct functionality tests. The OBC1 engineering model is based on the Analog Devices® PLUTO SDR Learning Module (ADALM-PLUTO). This device is similar in capabilities, SDR characteristics, and computing performance to the Sidekiq Z2 SDR. In addition, Pluto SDR and Sidekiq Z2 SDR are based on the same firmware. All these similarities allow the implementation and testing to be executed in the engineering model, then move the implementation towards the actual flight model for further physical testing. OBC2 engineering module is based on a "Raspberry Pi" zero, meaning that has the same hardware as the flight model. In this case, software migration towards the spacecraft is direct.

There are some main key concepts to address in this case of study. First, the OBC system has the main task to change the operation modes in the spacecraft to complete the mission, for which communication is essential. Also, the OBC must gather and organize information from commands and the spacecraft. For this, this subsystem interfaces with all the components inside the satellite. Finally, the use of the engineering model sometimes cannot be the same device as the one in the flight model, but similar hardware can be used for functional testing.

4.1 Payload

Payload is defined as the total complement of equipment carried, in this case, by the CubeSat, which ensures the fulfillment of the mission's objective. Typically, payloads are the main reason missions fly and what drives the mission size, cost, and risk.

Payload Systems deals with not only the systems on board a spacecraft tasked with delivering mission objectives but also the supporting ground equipment and telecommunication systems through which spacecraft payloads are controlled and results communicated to mission control.[27]

To define our space payload is necessary to question what we are trying to achieve, that is the reason that makes mission payloads unique in details of implementation and objectives, but it may share some characteristics with missions with similar objectives, the most common types of payloads are described in Table 4.

The payload determines what the mission can achieve, but the design of the spacecraft is influenced by the size of the payload and restrictions in structural, thermal, control, communication, or pointing characteristics.[28]

To correctly define the payload, there are some basic steps to do so, such as:

1. Select payload objectives, this will be strongly related to the mission objectives.
2. Conduct subject trades, evaluate what payload interacts with or observes.
3. Develop the payload operations concept, define de data produced by the payload and get it in the right format.
4. Determine the required payload capability, the performance required of the equipment.
5. Identify payloads and their specifications.
6. Estimate candidate payload characteristics and costs.
7. Examine the alternatives and preliminary selection of payload combinations.

---

[27] The European Space Agency (2018), About Payload Systems. https://www.esa.int/Enabling_Support/Space_Engineering_Technology/About_Payload_Systems.

[28] Pillet V., Aparicio A., Sanchez F., Payload and Mission Definition in Space Sciences, Cambridge, UK, Cambridge University Press, 2005.

Table 4 Types of space payloads

Type	Description	Sample function
Communication	The main goal is to transfer information	• Broadcast TV • Telephone • Wireless Internet
Observation	Any observation that a spacecraft makes without contacting the object	• Hubble Telescope • Meteorology
Navigation	Provides information on position, velocity, and time	• GPS • Galileo
In Situ Science	Sample collection and evaluation	• Mars Rovers • Lunar rovers • Gravity measurements
Action at distance	Concepts such as space-based solar power, space-based lasers, or space-based radar	• Guidance • Military data
Human Spaceflight	Human payload or flight crew	• Apollo

8. Determine mission utility in the function of cost.
9. Provide a detailed definition of the selected payloads on the requirements for the rest of the system.
10. Document and iterate the baseline payload design.

In the case of the K'OTO nanosatellite, since the objective of the mission is the remote sensing of the Mexican territory, we looked for a way to ensure the capture and sending of satellite images to the ground station. With this in mind, the OBC2 was born, which will be taken as a case study describing the payload components and their use.

Since it is a redundancy subsystem that should oversee fulfilling the K'OTO mission if something goes wrong, we begin by considering the potential problems that the satellite could have, considering the following scenarios:

- Scenario 1: In case the telecommunication subsystem fails and won't be able to communicate with K'OTO, it is necessary to have another way to send commands and download the satellite images taken to fulfill the mission. So OBC2 must have a way to communicate with the ground station independent to the rest of the subsystems, also, there must be an internal communication protocol with the main computer, to transmit to the commands that arrive from the ground station.
- Scenario 2: Another error that could occur is that the satellite camera could present a malfunction, so it is necessary to have another option to fulfill the mission because we are talking about a redundancy system, the backup camera must be of a different model and have different communication protocols, to reduce the possibility that both cameras break down.
- Scenario 3: On the other hand, the worst-case scenario would happen if the main computer failed, we would lose control with all the other subsystems,

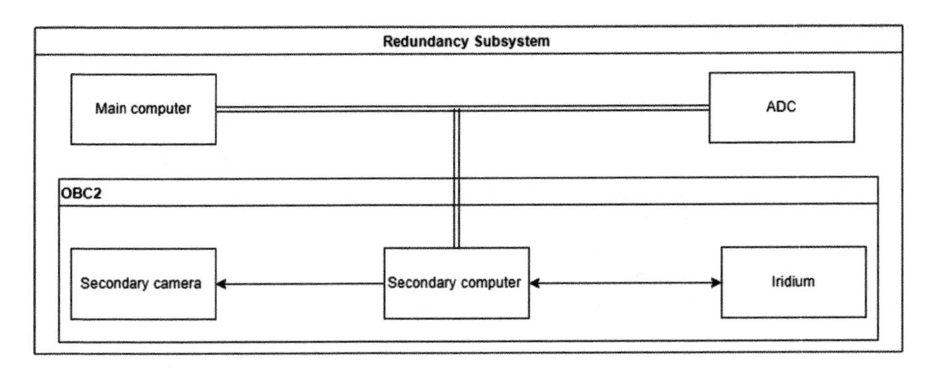

Fig. 24 Communication of redundancy subsystem of model K'OTO

therefore it is necessary that the OBC2 also has communication with the orientation control subsystem, called ADC, to be able to substitute the normal functions of the main computer.

Therefore, the objective of this subsystem is to take and send photographs to the ground station in case any of the main components fail, for this, the subsystem has the necessary components to be able to solve any failure previously mentioned.

Once the objective of the subsystem is defined, we can think about the payload that it should have, prioritizing that all the components of the OBC2 subsystem have flight heritage, that is, that they have been used in other missions. This is to reduce the probability of failure of any of them, in this case, we have a central computer, an Iridium communication module, and a second camera. In addition to having direct communication with the main computer and the ADC subsystem, this communication is described in Fig. 24, as well as the communication between its components.

Naturally, when working under the CubeSat standard, all components must be of low power consumption, low weight, and small dimensions, in addition to all the above, it is necessary to take special considerations for each component.

In the case of the camera, having two cameras, it was proposed to get a different lens than the main camera, so that the resulting images were different, likewise, the connection between the OBC2 and the secondary camera was different from the connection between the main computer and the main camera. Finally, we tried to make it possible to adjust the resolution and dimensions of the images, since these parameters directly affect the size of the image and this, in turn, affects the time it takes for the Iridium communication module of the OBC2 to send the image, thus increasing its energy consumption.

In the case of the communication module, the decision was made to use an Iridium module, since it is connected to the Iridium satellite constellation, the great advantage of this is that it is not necessary that the satellite is in line of sight or perform the pointing to communicate with it, since the system will identify the satellite in the constellation closest to K'OTO to communicate with him, because of this great advantage was considered appropriate OBC2, should not wait for the next orbit to read the error code in case of a failure or request a diagnosis.

Finally, the OBC2 computer must have enough computing power to replace the main computer, in addition to being able to handle different communication protocols, one to connect with the secondary camera, another for the Iridium communication module, and another for the main communication bus, although in this case, it must be in slave mode to connect with the main computer and in master mode to connect with the ADC subsystem. Another variable to consider is that it must be able to prioritize the tasks assigned to it, to have a short response time to be able to solve any error that may arise. Due to the above-mentioned, the processing power and memory of this computer should cover the demand of tasks that can be assigned to it.

Once defined the components are needed, it is necessary to define its operation, although the main objective of the subsystem depends on the occurrence of an unforeseen event, it must have a normal mode of use while the main subsystems work properly, this provides additional functions to the satellite by taking and sending photographs to the ground station under the control of the main computer, or by the instructions sent from the ground station. In this way, we can be sure that if any of the hypothetical cases occur, the OBC2 subsystem will be able to fulfill the K'OTO mission, but it will also be sending images of the Mexican territory even if the other subsystems are working normally.

To define the payload of a mission it is very important to have in mind the objective of the mission and the subsystem on it since this will determine the characteristics and the distribution of the components to be used in the mission, as well as the operation modes and the parameters that will determine if the payload is enough to carry out the mission or not.

5 Telecommunications

5.1 Telecommunications Subsystem Overview

The telecommunication subsystem is responsible for data transference between the satellite and one or multiple ground stations using radio waves. There are two types of links that may be established. The uplink is a communication link from a ground station to the satellite. It is normally used to send commands or data for retransmission and the downlink, which is a communication link from the nanosatellite to one or several ground stations. This depends on the hardware and software compatibility, only a receiver that meets the operational specifications of the transmission, such as data rate, modulation type, frequency spectrum, communication protocol, etc. Could receive the information transmitted. These parameters define the kind of hardware requirements, such as in the satellite, as in the ground station, like antenna type, gain, pointing system, transceiver, modem, decoder, power amplifier, and low noise amplifier.

5.2 Operational Requirements

This refers to the actions or status the system must perform to achieve the mission goal. These actions may be the time of operation of the radio receiver or transmitter, like missions with an always active periodical beacon signal, or if the transmitter is activated manually by a telecommand from a ground station. In either case, an operational requirement that is used in most satellites is to keep the radio receiver active by default to be able to receive a command at any time.

To define the communications subsystem requirements, it is important first to know what the nanosatellite mission parameters are, such as the type of orbit that can be seen in Fig. 25. This helps us to know the distances of the data transmission, the relative movement of the satellite from a ground station point of view to know the window time to establish a radio link with the spacecraft, to know these parameters is highly recommended to use prediction and simulation software such as Satellite Tool Kit (STK) to plan the mission. Figure 26 shows a simulation report generated for the K'OTO satellite that is a helpful tool to determine system requirements and match the results after performing the experiments.[29]

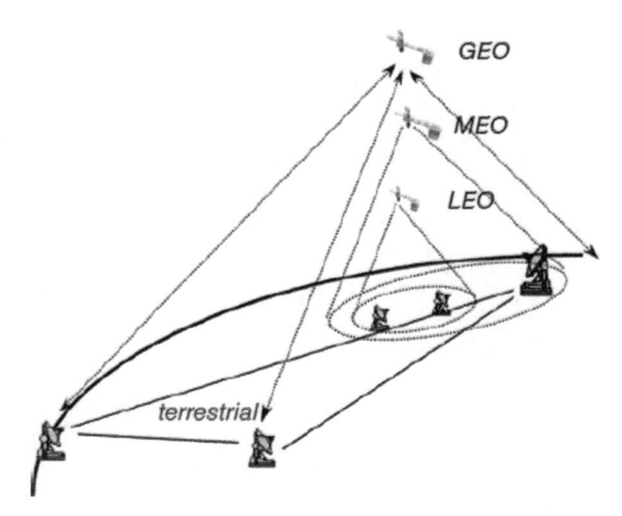

Fig. 25 Satellite sight range comparison between GEO, MEO, and LEO Orbits[30]

[29] Latachi, I., Karim, M., Hanafi, A., Rachidi, T., Khalayoun, A., Assem, N., … and Zouggar, S., Link budget analysis for a LEO CubeSat communication subsystem, 2017, International Conference on Advanced Technologies for signal and image processing (ATSIP), pp. 1–6. IEEE, May 2017.

[30] Gerard, Bousquet, and Zhili, Satellite Communications Systems: Systems, Techniques, and Technology (6th ed.), John Wiley and Sons Inc, 2020.

```
10 Jun 2020 20:44:57
Satellite-KOTO-To-Place-UAT:  Access Summary Report

KOTO-To-UAT
-----------
                     Access          Start Time (UTCG)              Stop Time (UTCG)
      Duration (min)
                     ------    ------------------------     ----------------------
--    --------------
                          1    1 Mar 2021 17:00:00.000      1 Mar 2021 17:02:49.9
79              2,833
                          2    1 Mar 2021 18:30:45.108      1 Mar 2021 18:36:41.9
30              5,947
                          3    2 Mar 2021 06:13:51.206      2 Mar 2021 06:23:29.2
33              9,634
                          4    2 Mar 2021 07:50:14.769      2 Mar 2021 08:00:13.0
87              9,972
                          5    2 Mar 2021 14:31:02.403      2 Mar 2021 14:31:40.5
25              0,635
                          6    2 Mar 2021 16:03:34.514      2 Mar 2021 16:14:01.7
91             10,455
                          7    2 Mar 2021 17:40:54.156      2 Mar 2021 17:49:35.6
60              8,692
                          8    3 Mar 2021 05:26:15.220      3 Mar 2021 05:34:06.1
75              7,849
                          9    3 Mar 2021 07:01:14.767      3 Mar 2021 07:11:52.4
55             10,628
```

Fig. 26 Simulation report of K'OTO satellite link access time obtained from STK

A key factor at the time of component selection for the communication on board subsystem depends on whether there is a ground station previously or if it will be built for the project specifically. In the first case, the component selection should be made around the ground station features to avoid incompatibility problems. In the second case, the ground station may be built or updated according to the project and the component selection may be made around the onboard subsystem. It is important to have in mind that it is better to invest more in the ground station than in the communication subsystem. It is not necessary to have a powerful transmitter onboard the satellite if there is a powerful receiver on the ground station. It's important to know the onboard radio transmitter is commonly the device with the most energy consumed on the satellite by far, and the power available on the satellite is a resource that must be handled with care, due to the amount of energy recovered by the power subsystem and its solar panels may not be enough for high-consume devices, this also determines the operation mode of the satellite, to know how often a transmission can be performed.

5.3 Design Requirements

The frequency band used may vary for each project. Most of the satellite missions use radio amateur frequencies that are allocated in a range for use exclusively for that purpose, which can be seen in Table 5. The use of each band is related to the amount of data to be downloaded. VHF and UHF bands are used with small dataset transmissions as Morse beacon signal or telemetry with payload information and S-band and further are used with higher datasets as high resolution.[31]

The bit error rate is the number of errors that may occur in data transmission. The threshold for digital communication links is 10E-5. For this value, less is better. A radio link with a BER (Bit Error Rate) of 10E-6 will not have major problems communicating information from an LEO orbit to a ground station. To choose the data modulation type as may be seen in Fig. 27 each one has factors to consider like power efficiency or bandwidth efficiency as listed in Table 6. The

Table 5 Frequency bands and their uses[32]

Band name	Frequency range (MHz)	Service	Direction
VHF	137–138	Space Operation Service Space Research Service	Download
	144–146	Amateur Satellite Service	Any
UHF	400,15–401	Space Operation Service Space Research Service	Download
	401–402	Space Operation Service Earth Exploration Service	Download
	402–403	Earth Exploration Service	Load
	410–420	Space Research Service	Inter Satellite Link
	430–440	Amateur Satellite Service	Any
	449,75–440	Space Operation Service	Load
S-Band	2.025–2.110	Space Operation Service Space Research Service Earth Exploration Service	Upload Inter Satellite link
	2.200–2.290	Space Operation Service Space Research Service Earth Exploration Service	Download Inter Satellite Link

[31] Barbarić, D., Vuković, J., and Babic, D., May, Link budget analysis for a proposed Cubesat Earth observation mission, in 2018 41st International Convention on Information and Communication Technology, Electronics and Microelectronics (MIPRO), pp. 0133–0138, IEEE, 2018.

[32] Atem de Carvalho, Estela, and Langer, Nanosatellites Space and Ground Technologies, Operations and Economics, pp. 115–142, John Wiley and Sons, 2020.

same modulation will not be the best candidate for every situation. This is related to the amount of the link's BER, because every modulation type will have different performance and maximum noise tolerance depending on the BER needed, as seen in Fig. 28.[33]

The frequency selection is also related to the kind of antenna to be used on board, and this may vary the coverage and a directivity requirement may be reflected in the type of ADCS system to establish a communication link. Some of the main antenna characteristics may be seen in Table 7. The line transmission design before the antenna is also important. Due to the transmission line impedance, all the communication systems must be driven with a general impedance of 50 Ω to ensure compatibility with other communication interfaces like connectors, transmitters, and transmission lines. This is important because a failure in this compatibility may cause a transmitter failure and mission loss. To measure the communication line impedance, a vectorial network analyzer may be used to find out the reactance of the system as shown in Fig. 29.

5.4 Legal Regulations

To operate a radio transmitter is important to achieve permission issued by the country's association affiliated with the International Amateur Radio Union (IARU) and the national communication institute. In Mexico, that institute is the Federal Telecommunications Institute (IFT) and the association representing radio

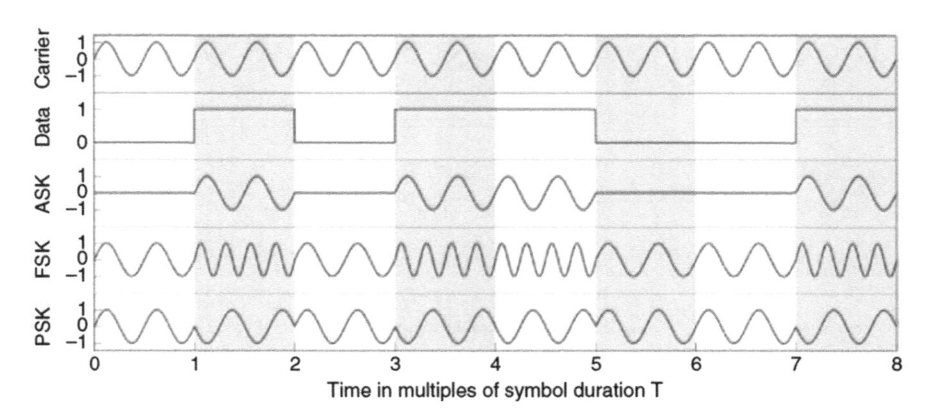

Fig. 27 Amplitude, Frequency, and Phase Modulation Waveforms Comparison when applied over Digital Data using a Carrier Sinusoidal Signal[34]

[33] Klofas B., 14 April 2016, CubeSat Communication System: 2003–2016, https://www.klofas.com/comm-table/table-20160414.pdf.

[34] Gerard, Bousquet and Zhili, Satellite Communications Systems: Systems, Techniques, and Technology (6th ed.), John Wiley and Sons Inc, 2020.

Table 6 Modulation types and characteristics[35]

Modulation	Modulation parameters	Bits per symbol	Bandwidth Efficiency	Sensibility to non-linear distortion
OOK (CW)	Amplitude (A)	1	High	Low
AFSK	Frequency (w)	1	Low	Low
FSK	Frequency (w)	1	Low	Medium
MSK	Frequency (w)	1	Medium	Medium
GMSK	Frequency (w)	1	High	Medium
BPSK	Phase (ϕ)	1	High	Low
QPSK/OQPSK	Phase (ϕ)	2	High	Medium
8-PSK	Phase (ϕ)	3	High	High
QAM	Phase (ϕ) Amplitude (A)	3^+	High	High

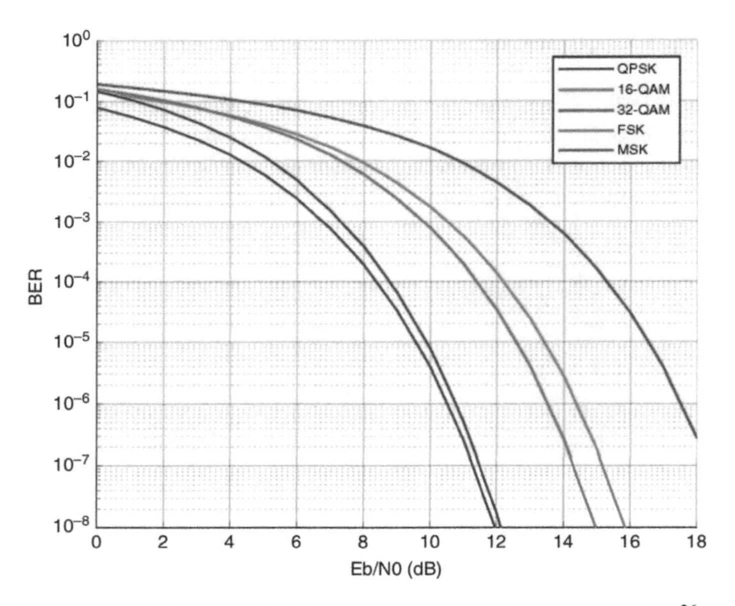

Fig. 28 Bit Error Rate Relation with Eb/No Requirement in different Modulations[36]

[35] Atem de Carvalho, R., Estela, J., and Langer, M., Nanosatellites Space and Ground Technologies, Operations and Economics,. pp. 115–142, John Wiley and Sons, 2020.

[36] Atem de Carvalho, Estela and Langer, Nanosatellites Space and Ground Technologies, Operations and Economics, pp. 115–142, John Wiley and Sons, 2020.

Table 7 On-Board Antenna Types and Typical Characteristics[37]

Antenna	Band	Gain	Directivity	Polarization	Volume	Mass
Monopole	VHF UHF	0–3 dBi	Omni	Linear or circular	10 mm × λ/4 (diameter)	<20 g
Dipole	VHF UHF	0–3 dBi	Omni	Linear or circular	10 mm × λ/2 (diameter)	<30 g
Patch	SHF XHF	5–8 dBi	60°	Linear or circular	10 × 10 × 0,1 cm (SHF) 5 × 5 × 0,1 cm (XHF)	<50 g

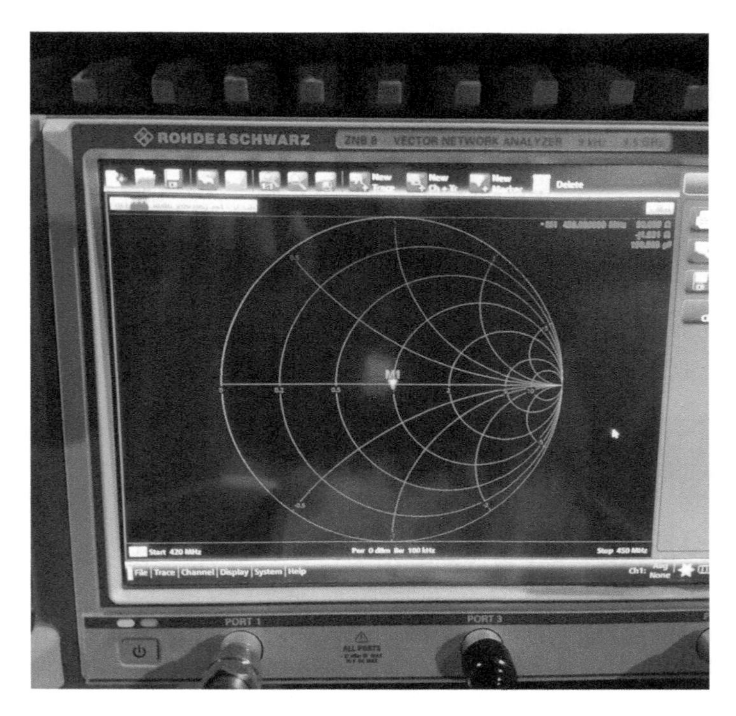

Fig. 29 K'OTO's Transmission Line Impedance Measurement using a Vectorial Network Analyzer

amateurs is the FMRE, A.C. as illustrated in Fig. 30. This permission is important to ensure the technical acknowledgment of radio transmission, ground station communication rules, and enforcement the law. This is permission that every ground station operator must have. Another of the most important permissions to obtain

[37] Cappelletti, Battistini, and Malphrus, Cubesat Handbook: From Mission Design to Operations, Academic Press, 2020.

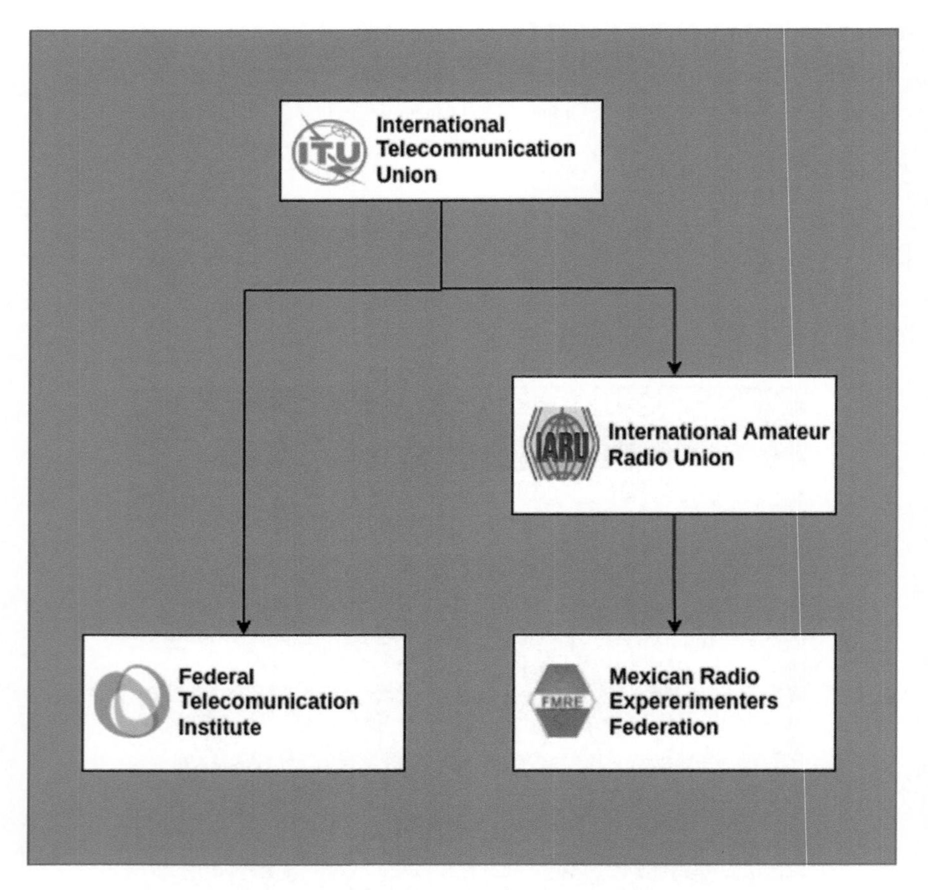

Fig. 30 The Organization of the Telecommunication Institutes Hierarchy in Mexico[38]

is the frequency allocation for the satellite operation. This may be the process with the longest wait queue, so it is important to start this process even before the planning stage of a satellite project. The permission is different depending on the spectrum that will be used to transmit the radio signals. If a satellite will be developed to use radio amateur frequencies, it is important to know, that nowadays, for satellite missions it is important to obtain licenses because the regulation of the spectrum is stricter than before and most of the accepted projects have a mission specifically focused on radio amateur service, other science payloads or experiments are not well seemed for the international committee.

[38] Adapted from AMSAT, https://amsat-uk.org/2017/06/30/iaru-satellite-coordination-wrc-15/, IARU aligns satellite coordination guidelines with ITU WRC-15 decisions, 30 June 2017, retrieved 9 January 2022.

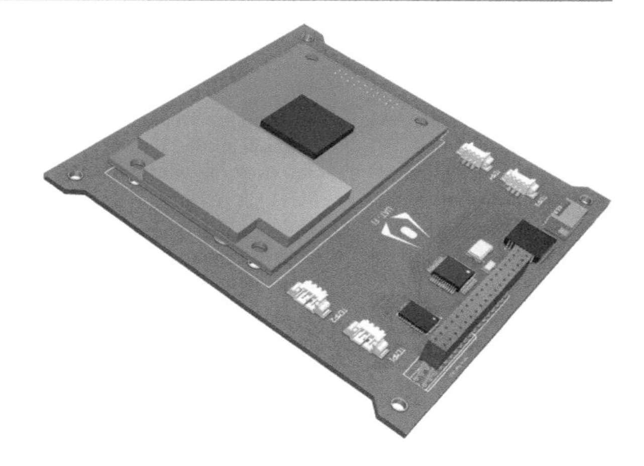

Fig. 31 ADCS board of K'OTO CubeSat (UNAM)

6 Attitude Determination and Control System

The Attitude Determination and Control System (ADCS) or ACS is defined as a system that determines and controls the attitude of a vehicle in space. The attitude (or the orientation) is always described concerning a reference system, either direction vectors within a reference coordinate system or as angles to defined reference axes.

The attitude of a spacecraft is independent of its position. Therefore, attitude control must be clearly distinguished from orbit control. Attitude control consists of a closed control loop with:

- Measurement and determination of actual attitude.
- Comparison with the desired attitude.
- Driving actuators through a dedicated controller to achieve the desired attitude.[39]

Figure 31 shows an example for an ADCS board. This system is capable of attitude determination from sensors in the board and placed in the external panels of the CubeSat, and to drive up to five magnetorquers.

6.1 Designing an Attitude Determination and Control System

The design and the type of the ADCS are determined by the functional requirements for the satellite. There are many methods for system designing. A list of

[39] Ley, W., Wittmann, K., and Hallmann, W., (Eds.), Handbook of space technology (Vol. 22), John Wiley and Sons, 2009.

Table 8 Steps in attitude system design

Step	Inputs	Outputs
(1a) Define control modes (1b) Define or derive system-level requirements by control mode	Mission requirements, mission profile, type of insertion for launch vehicle	List of different control modes during the mission Requirements and constraints
(2) Quantify disturbance environment	Spacecraft geometry, orbit, solar/magnetic models, mission profile	Values for torques from external and internal sources
(3) Select the type of spacecraft control by attitude control mode	Payload, thermal, and power needs Orbit, pointing direction Disturbance environment Accuracy requirements	Method for stabilization and control: three-axis, spinning, gravity gradient, etc.
(4) Select and size ADCS hardware	Spacecraft geometry and mass properties, required accuracy, orbit geometry, mission lifetime, space environment, pointing direction, and slew rates	Sensor suite: Earth, Sun, inertial, or other sensing devices Control actuators: reaction wheels, thrusters, magnetic torquers, etc. Data processing avionics, if any, or processing requirements for other subsystems or ground computer
(5) Define determination and control algorithms	Performance considerations (stabilization method(s), attitude knowledge and control accuracy, slew rates) balanced against system-level limitations (power and thermal needs, lifetime, jitter sensibility, spacecraft processor capability)	Algorithms and parameters for each determination and control mode, and logic for changing from one to another
(6) Iterate and document	All above	Refined mission and subsystem requirements More detailed ADCS design Subsystem and component specifications

steps in an iterative process is proposed by NASA for designing the ADCS. These steps are addressed in Table 8.[40]

[40] Starin, S.R. and Eterno, J., Aircraft Stability and Control, Attitude Determination and Control Systems, https://ntrs.nasa.gov/citations/20110007876, 2011.

Table 9 Stabilization Approaches for Various Missions

Mission	Pointing requirement	Maneuver	ADCS Type	Equipment	Relative Cost
Ram Pointing	3° into Ram	None	Aerodynamics at Altitudes below ~300 km only	Tail Feathers, Optical Pitch Bias Wheel, and Torque Rods with Magnetometer	Low
Nadir Pointing	>3°–10°	None	Gravity Gradient	Gravity Gradient Boom, Hysteresis Rods, No Attitude Sensing	Low
Nadir or Single Axis Inertial Pointing	1°	0° to 1° per day	Magnetic Spinner	Torque Rods, Magnetometer, optional Earth, or Sun Sensors	Low
Nadir Pointing	1°	Track Sub-satellite Point to Slight Pitch Axis Excursions	Pitch Momentum Bias	Single Reaction Wheel, Torque Rods, Magnetometer	Medium
3-axis Stabilized Earth or Inertial Pointing	<0,1°	None to Highly Agile	3-axis zero Momentum	3 Reaction Wheels, Torque Rods, Star Tracker(s), Magnetometer	High

6.2 Attitude Determination and Control System Requirements

There is a key question that helps to identify the main requirements when designing an ADCS: What are the pointing requirements for a satellite?

Most missions are either Earth or inertial pointing. Earth-pointing missions point a payload at or offset from, the sub-satellite point, requiring the spacecraft to rotate at or near the orbit rate. Inertial-pointing missions point the payload at the Sun or other target fixed in position to the stars. A kind of mission can have various agility requirements.

Mission requirements must flow down to specify the minimum pointing accuracy and spacecraft agility. Table 9 summarizes ADCS requirements for several types of missions.[41]

[41] Sebestyen, G. et al., (2018), Low Earth Orbit Satellite Design, Springer International Publishing.

6.3 Attitude Determination and Control System Architecture

The ADCS and every system can be thought of as having an architecture, but there are many different views or representations of that architecture that can be conceived, and these are described as follows:

- *The functional architecture:* a partially ordered list of activities or functions that are needed to accomplish the system's requirements.
- *The physical architecture:* at minimum a node-arc representation of physical resources and their interconnections.
- *The technical architecture*: an elaboration of the physical architecture that comprises a minimal set of rules governing the arrangement, interconnections, and interdependence of the elements, such that the system will achieve the requirements.
- *The dynamic operational architecture*: a description of how the elements operate and interact over time while achieving the goals.[42]

The physical and the technical architecture can be integrated into a single diagram since the physical diagram is the base for the technical one. Figure 32 illustrates the physical and technical relations between the ADCS components. The main component of the ADCS is the Computer/Processing Unit, where all data from sensors, devices, and communications are managed. This device can be an ASIC, an FPGA, or a microcontroller, depending on the application and the computational costs of algorithms, or the input/outputs requirements. Figure 33 shows an example of K'OTO ADCS architecture. In this system, some components have been chosen to satisfy the mission requirements.

The functional architecture of ADCS is useful for identifying the requirements and for constructing the model and simulation of the system in block diagrams. Moreover, it helps to identify the expected inputs and outputs for these functional blocks. Figure 34 shows a general architecture of a pointing ADCS. Note that this architecture is a closed loop system. The Body Dynamics inputs are the disturbance torque from Disturbances and the Initial Inputs data. The Attitude Determination inputs are the Inertial Models data and Body Dynamics state. The attitude determined is sent to Attitude Estimation which gives the estimated attitude to the Control Laws, which produces the Control commands to the Actuators. The Actuators produce the Control torque that again affects the body dynamics.

[42] Hastings, D., 16.892 J Space System Architecture and Design, Massachusetts Institute of Technology: MIT OpenCourseWare, https://ocw.mit.edu, License: Creative Commons BY-NC-SA, Fall 2004.

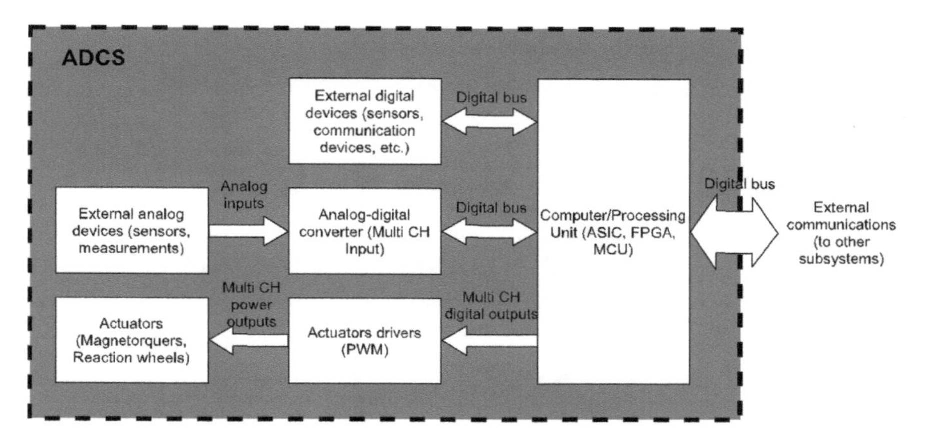

Fig. 32 Diagram of physical/technical architecture of an ADCS

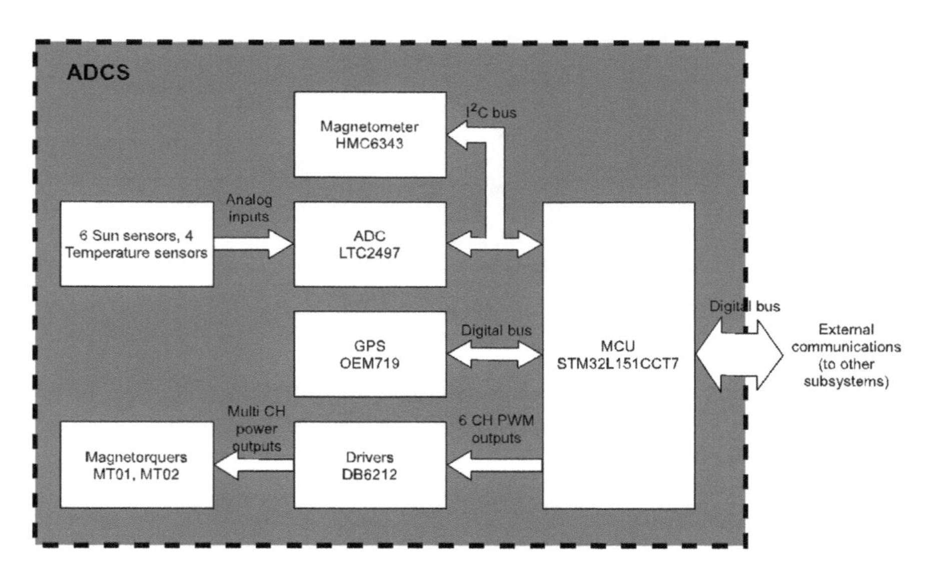

Fig. 33 Diagram of physical/technical architecture of the ADCS of K'OTO (UNAM)

6.4 Sensors Selection

Attitude sensors provide measurements of the actual attitude status. The attitude of the satellite is measured either as absolute or as relative to a certain frame of reference.

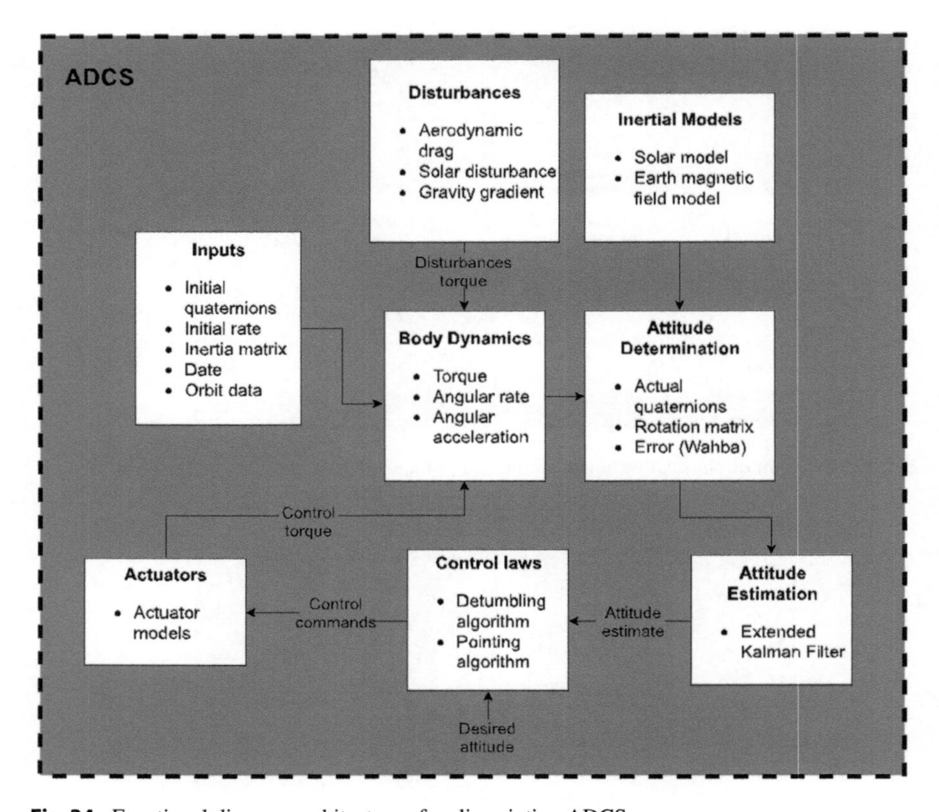

Fig. 34 Functional diagram architecture of nadir pointing ADCS

The vectors that can be measured on board are the following:

- Earth magnetic field vector.
- Direction vector to the Sun.
- Direction vector to stars.
- Direction vector to the Earth.
- Direction vectors to the satellites of a global navigation satellite system.

The accuracy of the three-axis attitude determination strongly depends on the accuracy of the single measurements and the relative orientation of the vectors.

6.5 Attitude Determination and Control System Testing

For validating the ADCS, some tests are to be performed. A peculiarity of such tests is to create conditions simulating space environments such as dynamic magnetic field, Sun radiation, and zero-gravity. The test bench must be equipped either with the satellite or with separated ADCS devices.

Preflight ADCS testing to verify performance reduces mission risk and helps operators and engineers better understand the systems that enable accurate pointing.[43]

There are some techniques for simulation and testing the ADCS. The most used technique is Hardware-In-the-Loop (HIL) simulation. Real hardware is included in the simulation loop and consists typically of sensors and/or actuators. HIL is a hybrid software/hardware simulation architecture, in which the hardware part can vary from a few pieces to a fully integrated system. The HIL technique is particularly useful for the verification of all those elements that operate in special environments and conditions which are difficult to reproduce in a laboratory. It may help to detect unexpected behaviors and/or failures arising from the integration of the component in the global system.[44] Various hardware configurations have been addressed.[45,46]

Most of the ADCS testbeds involve an air bearing as one of the main components of the rotating table since it allows for reducing the friction between the moving CubeSat and the test bed fixed elements. In satellite testbeds, air bearings are chosen primarily because of reduced friction that allows free rotation of the structure containing the satellite and leads to realistic simulation of the satellite dynamics in space.[47]

6.6 Helmholtz Coils Cage

A useful laboratory technique for getting a uniform magnetic field is to use a pair of circular Helmholtz Coils on a common axis with equal currents flowing in the same sense. This Helmholtz cage design has two constraints. The first constraint is the cage must be capable of generating a magnetic field twice as strong as the

[43] Tapsawat, W., Sangpet, T., and Kuntanapreeda, S., Development of a hardware-in-loop attitude control simulator for a CubeSat satellite, in IOP Conference Series: Materials Science and Engineering (Vol. 297, No. 1, p. 012010), IOP Publishing, 2018.

[44] Utah State University, Space Dynamics Laboratory, Attitude, Determination, and Control System (ADCS) Testing, https://www.sdl.usu.edu/capabilities/testing-calibration/small-satellite-test/adcs-testing.

[45] Corpino, S., and Stesina, F., Verification of a CubeSat via hardware-in-the-loop simulation, IEEE Transactions on Aerospace and Electronic Systems, -50 (4), pp. 2807–2818, 2014.

[46] Mahanti, K., Hardware-in-the-loop simulation and testing of the ADCS of the Beyond Atlas CubeSat, 2021.

[47] Gavrilovich, I., Krut, S., Gouttefarde, M., Pierrot, F., and Dusseau, L., Innovative Approach to Use Air Bearings in Cubesat Ground Tests, in CubeSat Workshop, 2016.

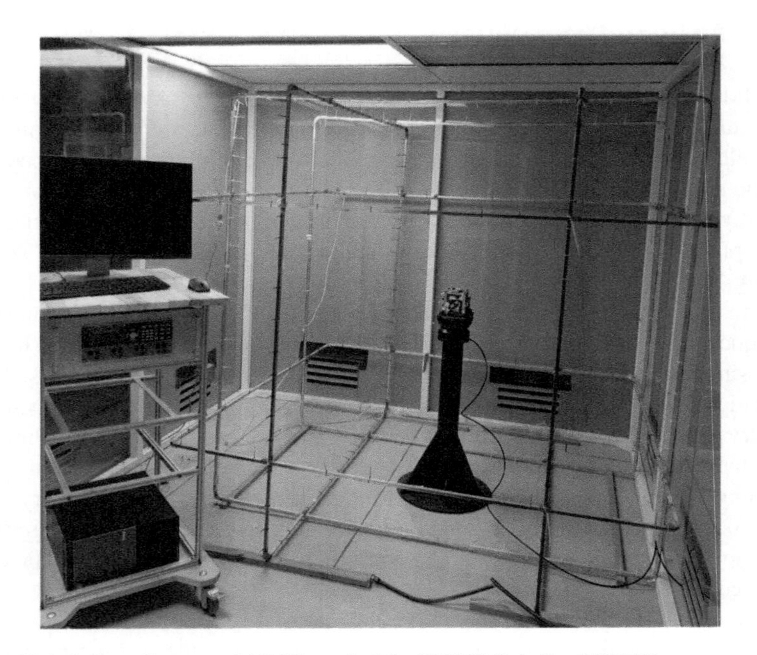

Fig. 35 Helmholtz coils cage of ADCS test bed for K'OTO CubeSat (UNAM)

Earth's magnetic field. The second constraint is that the field must be uniform over a large volume to fit a variety of CubeSats. Additionally, a Helmholtz cage must be entirely made from non-ferrous materials. Ferrous materials affect the magnetic fields, which would disturb the magnetic field produced in the Helmholtz cage.[48] Figure 35 shows the cage made for the tests for the ADCS of K'OTO CubeSat. The coils are each 1.80 × 1.80 m and have 1 m of separation at each reference axis. This Helmholtz cage can produce up to 1 Ga at its center in three directions.

6.7 Air Bearing Test Bed

A spherical air bearing test bed provides a frictionless testing environment. The bearing and load are supported by input compressed air. The compressed air creates a thin film below the half hemisphere and provides stiffness. Figure 35 shows an air bearing that is used to support a CubeSat structure.

[48] Stevens, J., (2016), CubeSAT ADCS Validation and Testing Apparatus, https://scholarworks.wmich.edu/cgi/viewcontent.cgi?article=3786&context=honors_theses.

7 Conclusions

This manuscript presented a methodology to fully develop a CubeSat. The steps mentioned here are the experience of the K'OTO team members through all the cycle life from the design until the final product delivery and that could find the best solutions and practices to have reliable results.

Human capital formation is a necessity for the development of the industry. These types of projects add practical experience to the student's basic university curricula. Also, the use of CubeSats in Mexico provides a good opportunity for enterprises for testing modern technologies in space. Even more, the university and associates acquire knowledge from this experience. Some benefits of this are boosting technological innovation, the enhancement future alike even outside of the academic environment, and networking for the participants.

It is important to point out that since 2003, the boom of the aerospace industry in Mexico began with the arrival of several prestigious companies which are leaders in the aerospace industry around the world. By 2022 it is expected to have 350 companies generating 52.000 jobs. It is projected that by 2024 to have 375 companies and 65.000 jobs, which is similar to conditions that were found in the country in 2019 before the Pandemic situation started.

This brief guideline for the development process of the K'OTO CubeSat offers a study case example to illustrate the different stages, beginning with the project lifecycle and an approach in the design process of the subsystems that conforms to the entire system. In practice, this kind of development does not show the product evolution along. Nevertheless, the interested reader and beginners may find this guide interesting to conceive a CubeSat project "from scratch".

Finally, it is important to remember that in a CubeSat development project, time is a crucial resource everyone must manage the best way possible. Due to this, it is necessary to know how long the processes the mission will need during operation, design, manufacture, and licensing to achieve a good project time planning and avoid problems during production due procedures that take more than a year like frequency licensing.

Dr. Rafael G. Chávez-Moreno is an associate professor at the National University Autonomous of Mexico-School of Engineering. He received his Ph.D. in Mechanical Engineering from the School of Engineering-UNAM. He is part of the Aerospace Engineering Department and is responsible for the Model Based on Design lab which belongs to the Space and Automotive Engineering National Laboratory located at Juriquilla. He is an active member of the Mexican Society of Mechanical Engineering and the Space Science and Technology Network. His current research areas include space systems, embedded systems, and control systems.

Dr. Jorge A. Ferrer-Pérez is an associate professor at the National University Autonomous of Mexico-School of Engineering. He received his Ph.D. in Aerospace and Mechanical Engineering from the University of Notre Dame, South Bend in the United States. He is part of the Aerospace Engineering Department and is responsible for the Space Propulsion and Thermo-vacuum lab. This facility belongs to the Space and Automotive Engineering National Laboratory located at

Juriquilla. His current research areas are nano-heat transfer in solid state devices, thermal control, space propulsion, small satellites, and the development of space technology.

Dr. Carlos Romo-Fuentes is an associate professor at the National University Autonomous of Mexico-School of Engineering. He received his Ph.D. in Technical Sciences in the Design of Space Systems considering electromagnetic compatibility criteria from the Aviation Institute of Moscow, Russia. He is part of the Aerospace Engineering Department and is responsible for the Electromagnetic Compatibility Laboratory. His current research areas are electromagnetic compatibility, certification tests, space systems, and space technology development. Likewise, is the technical responsible for the Space Science and Technology Theme Network from the National Council of Science and Technology from the Government of Mexico.

Dr. José A. Ramírez-Aguilar is an assistant professor at the National Autonomous University of Mexico School of Engineering. He received his Ph.D. in Technical Sciences in Radio receivers and microsatellites from the Moscow Aviation Institute-MAI, Russian Federation. He is the head of the Aerospace Engineering Department, responsible for the Ground Station Laboratory, and Vice-chair of the GRULAC of the International Astronautical Federation-IAF. His current research areas are Radio Frequency and microwave Systems, GNSS, Antennas, TT&C, Nano, and Microsatellites. Likewise, 2020 was selected for the first Latin American manned space mission ESAA-01EX SOMINUS AD ASTRA.

M.Eng. Sergio Ríos-Rabadán graduated from the Faculty of Chemical Sciences and Engineering of the UAEM as a mechanical engineer with postgraduate studies at the Faculty of Engineering of the UNAM with a master's degree in mechanical design in the satellite area, he also made a stay at the Kyushu Institute of Technology in Japan during his master's degree to complement his research thesis. He is currently collaborating with the K'OTO team in reviewing and complying with the design of the structure, as well as reviewing the mechanical environment of vibrations through FEM simulation and laboratory tests in such a way that the survival of the elements can be documented. subsystems based on the international space and military standards required by JAXA.

M.Eng. María G. Ortega-Ontiveros graduated from the Mechatronics Engineering career at the National Technological Institute of Mexico, Querétaro Campus, with postgraduate studies at the Faculty of Engineering of the UNAM with a master's degree in mechanical design in the satellite area. She is currently collaborating in the K'OTO structure design subsystem.

Xochitl Silvestre-Gutiérrez graduated from the Technological Institute of Durango with a bachelor's degree in computer systems engineering in data science. In 2019 she co-founded the scientific-technological association of Cazadores de Estrellas Durango A.C. She is currently collaborating in the documentation division within the K'OTO project, hosted at the High Technology Unit (UAT-UNAM) located in Querétaro, Mexico.

Eduardo Muñoz-Arredondo graduated from the Mechatronics Engineering career at the National Technological Institute of Mexico, Querétaro Campus. He worked from the 5th semester of the degree, as part of his social service, on the prototype that allowed the K'OTO nanosatellite to take the next step towards the beginning of its formal development. He is currently in charge of the K'OTO orientation control system, a system that is being developed from scratch at UNAM.

Saúl Zamora-Hernández Graduated from the Faculty of Informatics of the Autonomous University of Querétaro. He worked on his thesis about the communication subsystem design of the CubeSat KuauhtliSat (Ulysses II) at the UNAM high-tech unit in Queretaro. He is currently in charge of the K'OTO communication subsystem and ground station assistant.

Edgar I. Chávez-Aparicio is currently a student in the computer science master's program held at the Mathematical Research Centre (CIMAT), located in Guanajuato, Mexico. He received his Bachelor of Technology degree at the Applied Physics and Advanced Technology Centre (CFATA-UNAM). He has participated in the nanosatellite project "Ulises II KuautliSat." He is currently working in the OBC subsystem for the K'OTO project. Both nanosatellites have been hosted at the High Technology Unit (UAT-UNAM) located in Queretaro, México.

Saúl O. Pérez-Elizondo graduated from Technology career at the Applied Physics and Advanced Technology Center, UNAM Juriquilla. He participated in the design and implementation of the backup onboard computer subsystem of the K'OTO nanosatellite. He is passionate about the world of programming for the thrill of discovering new things constantly and is determined to specialize in the latest technologies in software development.

MSc. Bryanda G. Reyes-Tesillo is a Chemical Industrial Engineer born in Mexico City. She received his Master of Materials Science in the line investigation of polymers materials from the Universidad Nacional Autónoma de México. Her research is about coatings hybrid for aerospace applications.

Dr. Dafne Gaviria-Arcila is a Research Fellow at the Advanced Technology Unit of the National Autonomous University of Mexico. She has been involved in thermal and fluids analysis for more than 7 years. She was awarded 2020 her Ph.D. from the University of Nottingham in England. As part of her Ph.D. project, she was a recipient of the ZONTA International, Amelia Earhart Fellowship in 2017. Because of this recognition, in 2018 the Honorable Mention of Carlos Fuentes Award from The Mexican Embassy in the United Kingdom. She has participated in different activities to influence women to study STEM careers such as mentoring, inspiring, and empowering them. Her story has been featured in the 100 extraordinary Mexicans edition of the book titled Good Night Stories for Rebel Girls.

Printed by Printforce, the Netherlands